A-Z NANOBIOLOGY

Albert Shawn

CENTRUM PRESS
NEW DELHI-110002 (INDIA)

CENTRUM PRESS

H.O.: 4360/4, Ansari Road, Daryaganj,
New Delhi-110 002 (India)
Ph.: 23278000, 23261597

B.O.: No. 1015, Ist Main Road, BSK IIIrd Stage
IIIrd Phase, IIIrd Block,
Bangalore - 560 085 (India)
Tel.: 080-41723429
Visit us at: www.centrumpress.com

A-Z Nanobiology

First Edition, 2009

ISBN 978-93-80106-47-2

PRINTED IN INDIA

Printed at Salasar Imaging Systems, Delhi-110035 (India)

Contents

	Preface	*vii*
1.	Nanobiology	1
2.	Nanoorganisms	23
3.	Tools in Nanobiology	77
4.	Nanobiology Strategies	86
5.	DNA Microarray	95
6.	Hemicellulolases in the Nanobiology	107
7.	Future of Stem Cells	134
8.	Disease Transmission	150
9.	Uses of Microbes	246
10.	Tissue Culture and Laser Application	291
	Index	303

Preface

Biological systems are inherently nano in scale. Unlike nanotechnology, nanobiology is characterized by the interplay between physics, materials science, synthetic organic chemistry, engineering and biology. Nanobiology is a new discipline, with the potential of revolutionizing medicine: it combines the tools, ideas and materials of nanoscience and biology; it addresses biological problems that can be studied and solved by nanotechnology; it devises ways to construct molecular devices using biomacromolecules; and it attempts to build molecular machines utilizing concepts seen in nature. Its ultimate aim is to be able to predictably manipulate these, tailoring them to specified needs.

Nanobiology targets biological systems and uses biomacromolecules. Hence, on the one hand, nanobiology is seemingly constrained in its scope as compared to general nanotechnology. Yet the amazing intricacy of biological systems, their complexity, and the richness of the shapes and properties provided by the biological polymers, enrich nanobiology. Targeting biological systems entails comprehension of how they work and the ability to use their components in design. From the physical standpoint, ultimately, if we are to understand biology we need to learn how to apply physical principles to figure out how these systems actually work. The goal of nanobiology is to assist in probing these systems at the appropriate length scale, heralding a new era in the biological, physical and chemical sciences.

Author

Preface

Chapter 1

Nanobiology

Biological systems are inherently nano in scale. Unlike nanotechnology, nanobiology is characterized by the interplay between physics, materials science, synthetic organic chemistry, engineering and biology. Nanobiology is a new discipline, with the potential of revolutionizing medicine: it combines the tools, ideas and materials of nanoscience and biology; it addresses biological problems that can be studied and solved by nanotechnology; it devises ways to construct molecular devices using biomacromolecules; and it attempts to build molecular machines utilizing concepts seen in nature. Its ultimate aim is to be able to predictably manipulate these, tailoring them to specified needs.

Nanobiology targets biological systems and uses biomacromolecules. Hence, on the one hand, nanobiology is seemingly constrained in its scope as compared to general nanotechnology. Yet the amazing intricacy of biological systems, their complexity, and the richness of the shapes and properties provided by the biological polymers, enrich nanobiology.

Targeting biological systems entails comprehension of how they work and the ability to use their components in design. From the physical standpoint, ultimately, if we are to understand biology we need to learn how to apply physical principles to figure out how these systems actually work.

The goal of nanobiology is to assist in probing these systems at the appropriate length scale, heralding a new era in the biological, physical and chemical sciences. Biology is increasingly asking quantitative questions. Quantitation is essential if we are to understand how the cell works, and the details of its regulation. The physical sciences provide tools and strategies to obtain accurate measurements and simulate the information to allow comprehension of the processes. Nanobiology is at the interface of the physical and the biological sciences. Biology offers to the physical sciences fascinating problems, sophisticated systems and a rich repertoire of shapes and materials.

Inspection of the protein structure databank illustrates the breadth of scaffolds, shapes and properties that protein molecules and their building blocks can provide. Via a shape-guided self-assembly strategy, these can be put together toward a specific function. Further, by inserting synthetic non-natural residues at judiciously selected positions, or synthetic peptide linkers, we may selectively rigidify the construct, or obtain a totally new world of shapes and scaffolds. Such broadening of the chemical space may lead to an almost unlimited range of nanosystems and architectures. Merging computation with experiment will accelerate nanodesign.

Computational modeling will enhance the application of nanotechnology to key areas such as drug delivery and biomaterial design. Nanobiology is a field where interdisciplinary collaborations are essential and disciplines converge. Discipline convergence should enable the quantitation, leading to a better understanding of the regulatory networks within cells and between cells of an organism.

These networks dictate how a cell responds to external stimuli, which in turn activate signaling cascades. It should allow the addressing of a broad range of questions on the structure and function of the cytoskeleton; the nuclear envelope; signal transduction by membrane embedded receptors; the nanomechanical properties of the extracellular matrix; nuclear transport; and voltage induced channel gating.

For successful nanostructure design, we need to figure out and be able to control the intermolecular associations. For a stable functional construct, there are two key elements:

first, the conformations of the building blocks in the designed structure should follow their natural tendencies; and second, the associations should be favorable. Molecules interact through their surfaces. Thus, favorable associations derive from shape complementarity and contributions of the various physical components.

Nanobiology is in its infancy. Yet, biology provides an enormous range of engaging and stimulating problems with many *in vivo* examples of intricate, complex, fascinating biological systems. Understanding, mimicking and controlling the devices which target these processes and which are constructed from these molecules is a tremendous challenge to the converging disciplines in nanobiology.

NANOMACHINES

Having devised nanomachines capable of manipulating single atoms and molecules and begun to apply these technofeats to modify the material world around us, scientists are now shifting their gaze inward. to the human body. They envision nanorobots coursing through our bodies able to fix dangerous mutations on the spot. cleanse blood vessels or secrete insulin to counter an abnormal rise in blood sugar levels.

The quest of nanomedicine, however, has had its fair share of controversial enthusiasts, offering grandiose scenarios of future medical therapies. Set in motion by Richard Feynman's 1959 lecture on miniaturization, it captured popular imagination and picked up speed following Isaac Asimov's Fantastic Voyage. This science fiction classic describes a crew of miniaturized humans rushing to save Professor Beans, a miniaturization expert who has suffered a life-threatening blood clot deep in his brain. The rescue team navigates through Beans' circulatory system in an atomic-scale submarine.

At the time, most of his ideas were deemed far-fetched at best, but now, nearly 20 years later, prototypes of his proposed scenarios are beginning to pop up in research labs around the world. At the Weizmann Institute of Science, researchers have created a tiny computer built of DNA molecules that detects prostate cancer conditions in a test tube. And elsewhere in the nanoscience world, researchers are working to create soap-like films to encapsulate drugs that could deliver their medicinal cargo right on target.

The films would protect the drug in the bloodstream, releasing it only under certain chemical or thermal conditions, thus preventing wholesale release of a potentially toxic drug. Other nanotools may identify the presence of bacteria and viruses, which, like many biological particles, have welldefined electrical properties and thus oscillate in characteristic ways when exposed to an electric field. Numerous challenges remain – technical, scientific and even psychological – until nanomedicine will directly influence healthcare. Here's a look at some of the early steps taken here at the Institute.

LEARNING FROM NATURE

In the early 1940s, Swiss engineer Georges de Mestral went for a walk in the woods with his dog. Upon his return he noticed they were both covered with seeds of a burdock plant. Removing the burs proved a frustrating task. Taking a closer look, de Mestral observed that the burs had hundreds of tiny little hooks – a clever trick for a plant wishing to disperse its seeds. He thought this might be an ideal means of joining two fabrics together – and the idea of Velcro was born.

Today Velcro is found on everything from zippers to sneakers to space suits, and scientists are working on yet another form that rips apart silently – a critical ingredient for military operations. Velcro is a classic example in the story of biomimetics, a field that has scientists turning to nature to examine its countless bioengineering feats, polished over time.

The hope is that new ideas for advances in medicine, technology and industry might be hidden in some of the solutions evolved to meet the challenges of life – from bones and teeth, capable of withstanding decades of grinding wear and tear, to the enormous tensile strength and elasticity of spider dragline silk, which ounce for ounce is stronger than steel, enabling a dangling spider to safely dive down and nab its prey.

Dealing with the daily grind

A team of Institute researchers has now discovered a key factor explaining this hardiness: a spring-like structure that absorbs the mechanical stress incurred as teeth chew, grind or, as circumstance would have it, chatter their way through an arctic freeze. Interestingly, this structural feature constitutes only a tiny percentage of the overall tooth.

The team, consisting of Prof. Stephen Weiner of the Structural Biology Department and post-doc Dr. Rhizi Wang, began by scouring the literature in search of a dental component that might fit the bill of mechanical stress absorber. "Drawing on his materials science expertise, Wang put his finger on the core question," says Weiner. "Given that the two main components of teeth are enamel – a hard, brittle material – and dentin, which is much softer, Wang hypothesized that there must be an intervening layer that allows these two very different materials to work together."

The researchers knew that studies dating back to the 1960s had uncovered a certain structural zone positioned directly between enamel and dentin, but they found no follow-up studies exploring its significance. They decided to determine whether teeth actually compress preferentially in this unexplored area when subjected to mechanical stress.

As published in the *Journal of Biomechanics*, the team found that the structure in question, which they dubbed the "soft zone," is where almost all compression takes place. The study showed that the zone – which is merely 200 microns thick, compared to enamel's 1,000 microns and dentin's 5,000 – is in fact the "working part" of the tooth. When we chew, the enamel part of our teeth is pushed back while the soft zone compresses. It then functions like a spring, bouncing back after the compression.

Biological computer diagnoses cancer

The team programmed their computer to detect prostate cancer and one form of lung cancer. The computer evaluates four genes that become either under-or overactive once the disease sets in. The genes chosen control the expression of messenger-RNA (which carries information from the nucleus to the ribosome, the cell's protein factory). The scientists introduced different levels of these RNA molecules into the test tube to simulate the presence or absence of cancer. Made entirely of biological molecules, the computer has three components _ input, computation and output. The first consists of short strands of DNA, called transition molecules, which check for the presence of the messenger RNA produced by each of the four cancerous genes.

The second component is a computation (diagnostic) module, consisting of a hairpin-shaped long DNA strand. As the computer's input component checks for the presence or absence of the four cancerous markers, this diagnostic unit checks each input in turn, producing a positive diagnosis of malignancy only if all four markers point to cancer.

This diagnostic module also contains the computer's third component: a single-stranded DNA that is known to interfere with the cancer cell's activities. In the case of a positive diagnosis, the unit releases its hold on the therapeutic unit, activating its cancer-fighting potential. Shapiro's team first went on record in 2001, when it created the first autonomous biological nanocomputer. Made entirely of biological molecules, the computer's input, output, and "software" were made up of DNA molecules.

The device was so small that as many as a trillion such computing devices could work in parallel in one drop of water, collectively performing a billion operations per second with greater than 99.8% accuracy per operation. It was recently awarded the Guinness World Record for the "smallest biological computing device." Shapiro: "Our study offers a vision of the future of medicine. It is clear however, that it may take decades before such a system operating inside the human body becomes a reality."

THE SLICK JOINT

Mimicking key design elements of this biolubrication system, physicist Prof. Jacob Klein of the Weizmann Institute of Science, has recently created a synthetic lubricant that cuts friction by a thousand-fold or more. The study, published in *Nature,* could lead to a range of applications – from longer-lasting micromachines and higher-density hard disks to biomedical devices, including improved hip transplants and treatment for dry eye syndrome.

Previous studies had suggested that biolubrication systems, including those in our joints and eyes, may contain hyaluronan, a molecule that coats the rubbing surfaces of our joints, shielding them from mechanical damage. Hyaluronan was also known to be strongly attracted to water. But how these two factors combined to create the most effective lubricatory system anywhere, remained a mystery. Klein and colleagues suspected that in joints, hyaluronan may be attached to a thin cartilage layer covering the bone, while parts of this long, chain-like molecule stick out into the synovial fluid between the bones, similar to bristles on a hairbrush. These bristles or others playing a similar role, they believed, function as the body's lubricants.

To test their theory, the team developed a synthetic model that mimicked a double-brush system, anchoring two charged molecules (polyelectrolytes) to oppositefacing ceramic surfaces. The resulting system showed extremely effective friction resistance, particularly when exposed to a water-based solution. "The brushes strongly try to avoid each other, resisting penetration even when an external force is applied that presses them closer. This enables them to easily slide past one another," says Klein.

The superior lubrication may largely be explained by electrical charges. The synthetic brushes were designed to imitate the electrically-charged nature of biolubricants. Similar to the negative charge characterizing the hyaluronan molecule, for instance, the end of the synthetic brushes stretching away from the ceramic surface was designed to be negatively charged.

This negative charge then attracted water molecules – which in fact explains why the brushes performed most effectively in a water-based solution. "The water molecules are tightly bound to charges on the brushes, causing them to act like molecular ball bearings to reduce friction," says Klein.

Other players in this charge cast are the small mobile ions trapped in the space between the bristles. According to Klein, when the brushes approach one another, their respective clouds of positive ions are mutually repellant, increasing the brushes "distaste" of penetrating one another. "Our general idea was to draw on nature, says Klein. "It made sense to try out the design principles evolved and optimized over millions of years."

HUMPTY-DUMPTY MECHANICS

How bones withstand the immense array of mechanical stresses imposed upon them has long fascinated scientists, who have set about probing their chemical and biological properties and even crushing them to determine the relationship between their structure and their outstanding performance. Yet despite several decades of research, a full understanding of this complex biological material remains elusive.

The beauty of bone stems from two key properties – it is a composite material and it is highly hierarchical in structure. The principle of forming a composite is to take two materials, often highly different from each other – say, a soft and an extremely hard material – weave them together somehow, and, presto, a new material is born that is generally far superior to either of its individual components.

In bone, this partnership comes in the form of a ceramiclike, brittle mineral salt called hydroxyapatite, which is bolstered by collagen (a protein) that adds elasticity and strength. The tissue has several levels of organization, ranging in scale from nanometers to centimeters, each with a different structure and function. Adding to this complexity is the fact that bone changes over time, with specialized bone cells actively replacing older bone tissue in a process that has adults renewing their entire skeleton roughly every 7-10 years.

In the 1990s Prof. Stephen Weiner of the Institute's Structural Biology Department and Prof. Daniel Wagner of the Materials and Interfaces Department collected data on the micro-mechanical properties of bone and developed a mathematical model explaining how bone's various components affect its mechanical function, including bone elasticity. The model made it possible to predict the mechanical properties of bone with unprecedented accuracy.

Weiner is currently working to further understand the contribution to mechanical strength made by the individual cylinders within bone. Each of these cylinders changes over time, altering its mineral content and thus its mechanical properties. "While past studies succeeded in determining the average mechanical properties of bones, what we really need to know is how each of these individual cylinders responds to stress," says Weiner. Insights at this structural level may yield new treatments for osteoporosis and other bone diseases.

FINE SPINE CHOICES

Scientists had long been unable to explain the durability of sea urchin spines, which they believed consisted entirely of a single calcite (and thus highly fragile) crystal. The Weizmann scientists revealed that while calcite indeed constitutes 99.9 per cent of the urchin spine, the remaining 0.1 per cent consists of proteins entrapped within the crystalline matrix.

And amazingly, that is what makes all the difference. Working with researchers from the Brookhaven National Laboratory in New York, the Weizmann team discovered how the entrapped proteins act to reinforce urchin calcite against fracture, preventing cracks from spreading through the crystal. The scientists grew pure calcite crystals in solutions containing proteins extracted from urchin spines. They found that the proteins integrated into the crystal, altering its structure and making it far less brittle.

The question remained, however, how a component present in such minuscule proportions could have such far-reaching effects on material quality. Follow-up studies showed that the proteins had integrated into the pure crystals preferentially, along planes that are oblique to the crystal cleavage plane, thus disrupting fracture propagation. "The crystal's ability to withstand mechanical stress is enhanced because resulting cracks do not rip catastrophically along the cleavage planes but are diverted along the protein-induced planes, which absorb the force of impact," explains Weiner. These findings might lead to the development of new crystal-polymer composites for building lightweight, tougher ceramics and improved electrooptic materials.

"Sightless" Marine Creature Found to be All Eyes

Profs. Lia Addadi and Stephen Weiner of the Weizmann Institute's Structural Biology Department had long been interested in the ways in which animals build their skeletal structures. When they met Dr. Gordon Hendler of the Natural History Museum of Los Angeles County, Hendler brought to their attention one particular species of brittlestar that appeared to be particularly sensitive to changes in light, quickly escaping into dark crevices at the first sign of danger. He suspected that spherical crystal structures on the brittlestar's outer skeleton serve as lenses, transmitting light to its nervous system.

By analyzing the geometry of the crystal lenses Addadi and Weiner, together with their then graduate student Joanna Aizenberg, were able to pinpoint the expected focal point on the nerve bundles below, but they lacked the means of proving that these lenses indeed transmit light to the nervous system within.

This is where things stood for almost ten years, until the team came up with the idea of examining the lenses using lithography, a semiconductor technology. Placing one of the crystals above a layer of photosensitive material, Aizenberg exposed the system to light and found that it reached the photosensitive tissue in spots directly underneath the crystals. Her findings also demonstrated that the crystalline lenses act as "corrective glasses," filtering and focusing light on photoreceptors within the nerves.

But unlike man's ability to see in virtually only one direction, this complex visual system enables the brittlestar to detect approaching danger from many directions. The lenses expertly compensate for common optical distortion effects. This unique visual architecture has prompted hopes for new materials that would mimic the brittlestar model. Knowing how to build such microlenses of high optical quality could lead to improved microlithography tools used in etching the integrated circuits found in computers.

Peering into Nature's Secrets

Cells vary in length. Bacteria are about 1 micron long, while plant and animal cells vary from about 50 microns to meters in the case of certain neurons.

The cell – the elementary unit of life – is rich in engineering know-how. The ultimate self-assembler, it applies its molecular machinery to convert readily available materials from food and water into a wonderland of materials – from over 100,000 proteins, lipids and sugars, to the materials they build: wood, bones, horns, scales and skin. It performs these feats in a tiny setting. Cells are generally only a few micrometers long, and their functional units, the organelles, are only a few nanometers.

Scientists around the world are hard at work applying a range of tools to chip away at the cell's secrets by viewing and manipulating nanoscale objects within and around the cell. Lessons from the cell, they believe, might prove vital in understanding the basic mechanisms of life – the key to unprecedented medical advances. The cell might also help in shaping our material world, with its varied nanostructures serving up solutions to central challenges in material design.

Connect the Dots

Until recently, scientists studying the forces influencing cell adhesion faced a major difficulty: they lacked the tools to effectively identify and measure the minuscule forces applied between cells and their environment. Heading a multidisciplinary team, Prof. Benjamin Geiger of the Institute's Molecular Cell Biology Department developed a powerful new tool for studying cellular adhesion, which allows the measurement of forces in the Lilliputian range of nanonewtons (one nanonewton is roughly equivalent to the weight of one thousand red blood cells). The method involves creating a transparent elastic template imprinted with a grid of dots or lines spaced at fixed intervals.

When a cell sticks to the patterned template and applies a force, it distorts the grid. By measuring these distortions, the scientists are able to infer the magnitude and direction of the forces applied by the cell to the contact area. This approach is proving vital in revealing the interplay between adhesion and internal cellular processes.

In related research, Geiger and Prof. Alexander Bershadsky of the Molecular Cell Biology Department have shown that cells constantly "probe" their surrounding by touching and pulling at it. New research questions include the nature of the forces occurring between two cells as they anchor on the same substrate, a topic that plays a role in how cells aggregate to make a tissue and may thus have relevance to tissue engineering.

Simulating life

Collaborating with Prof. Lia Addadi of the Institute's Structural Biology Department, Ph.D. student Baruch Zimmerman and Prof. Yoachim Spatz of Germany's Heidelberg University, Geiger is creating artificial micro- and nanoscale models that simulate the body's varying tissue matrices (collagen, fibers etc.) to obtain a better understanding of the interactions between a cell and its environment – specifically, how a cell's behaviour and motility are affected by the chemistry and mechanical properties of its environment.

Insights into these questions may have implications in cancer research aimed at preventing the migration of cancerous cells to form metastases. The artificial models tested at the WIS may also advance tissue engineering research, yielding knowledge of the type of scaffolds needed to create specific tissues.

Soft Matter Science

Also known as complex fluids, soft matter often deals with the dispersion of a solid or liquid in another liquid.

In his work at Exxon, Safran, today a professor in the Institute's Department of Materials and Interfaces, focused on unique "split-personality-like" molecules used in oil extraction and recovery. Known as amphiphiles, these molecules have both polar (charged) and nonpolar regions. While one end is attracted to polar molecules such as water, the other end is generally a nonpolar hydrocarbon chain, attracted to oils or lipids. This enables amphiphilic molecules to break up oil slicks into small droplets, which might then be further degraded by oil-eating microbes.

The unique polar/nonpolar duality of amphiphiles makes them key players in the body, where they selfassemble into an extremely rich assortment of soft matter structures that perform regulatory and housekeeping chores.

For instance, the membranes of all cells in the body contain a nonpolar region facing the inside of the cell and a polar region on its surface, which is vital to the cell's interaction with nearby molecules. Another example is that of micelles – a cluster of molecules critical to efficient digestion, which break up otherwise insoluble fat molecules. A better understanding of the body's self-assembling systems may help in creating vesicles for use as potent drug delivery systems.

Opening the Gates

These gates (called nuclear pore complexes) lead to and from the cell's genetic material, safeguarding the essential codes of life. Only "desirable" molecules can pass through. Molecules that succeed in entering the nucleus from other parts of the cell activate genes, causing the information these genes encode to be dispatched to the cell's "protein factory" by another molecule – messenger RNA – which exits the nucleus through the same gates.

Understanding the mechanism by which molecules pass the nucleus' gates is crucial to controlling such traffic. These methods may be applicable in barring the entry of viruses into the cell's nucleus, or in gene therapy, in which a beneficial gene is introduced into the patient's DNA to replace a damaged counterpart gene. Dr. Michael Elbaum of the Weizmann

Institute's Material and Interfaces Department is studying the passage of DNA molecules through these gates using optical and electron microscopy. His team applies a tool based on tiny tweezers composed of laser beams focused in a microscope.

In collaborative research with Prof. Alexander Bershadsky of the Molecular Cell Biology department, Elbaum and his team applied similar optical tweezers to probe cell adhesion and its relation to cell movement in the body – phenomena playing a key role in embryonic development, wound healing and other fundamental life processes. The team designed small beads coated with proteins involved in adhesion.

They then used the optical tweezers to hold the beads over the cell surface and map their resulting interaction with the cell. In the study, published in *Biophysical Journal,* they showed striking differences in the way different regions of the cell surface respond to the stimuli generated by the adhesion proteins, making it possible to pinpoint specific regions where adhesion takes place.

Metal 'a Day to Keep Diseases Away

Prof. Daniella Goldfarb, of Institute's Chemical Physics Department, is studying metalloenzymes (enzymes containing iron and other metals) with the aim of mapping the precise structure of their metalcontaining active sites. Insights in this field will help advance the construction of molecular machines – 10 to 50 nanometers large – that, equipped with these metal-containing active sites, would be capable of performing diverse industrial functions, some dramatically improved from existing technologies.

A good example is that of ammonia production – a reaction that calls for extreme temperatures and pressure when performed industrially, yet is synthesized in bacteria in a series of elegant biochemical reactions, at ambient temperatures and without the production of harmful byproducts. Mapping the metal-containing sites may also lead to methods for repairing damaged enzymes, offering an invaluable medical tool since damaged or faulty enzymes lie at the root of many diseases.

First Sightings of Individual Proteins as They Fold

Until recently, scientists studying protein folding had to rely on information gathered from a huge number of molecules. The experimental results represented an average of the different processes these molecules underwent and could not pick up individual differences. This shortcoming highlighted the need for a new technology that would make possible the study of individual proteins as they fold – a huge challenge since proteins are generally only a few nanometers in size and are constantly on the go. Applying a new optical technology designed in his lab, Dr. Gilad Haran of the Institute's Chemical Physics Department has made the first glimpses ever of single proteins in the process of folding.

The technology's success lies in striking the right balance. Since proteins are continuously active, one must limit their motion to get a clear understanding of their folding process. However, a fully immobilized protein defeats the purpose. It won't fold. Earlier attempts to gain a steady look at proteins sought to pin them down to a surface; but this could be achieved only by binding the protein to a surface, thus changing its properties. To skirt this obstacle, Haran's team designed novel vesicles that envelop the proteins in question.

Each vesicle is 100 nanometers wide and designed to envelop a single protein molecule, generally only 3-4 nanometers long. Unaware of their scientist-made borders, the proteins can move about freely, yet not so actively as to impair the scientists' ability to observe their behaviour.

The results verify what theoretical scientists have suspected for nearly a decade – that proteins may vary in their folding process. Even identical proteins ending up with the same shape may take different routes to reach it. The new WIS technology might help clarify the reasons for protein misfolding and ensuing disease.

DNA and Proteins

AFM works like a record player "reading" surfaces with a needle-fine tip; it rises and descends as it meets "bumps" or "valleys" in the target surface. These motions are translated via a computer to create an atomic-scale image. But while AFM provides detailed topographical information, it offers only limited information about an object's surface chemistry. "It's like being able to determine where a cake is without knowing whether it's a chocolate swirl or blueberry with cream," says Joselevich, of the Institute's Materials and Interfaces Department.

Seeking to enhance the tool's sensitivity, Joselevich, at the time completing a post-doc at Harvard under Prof. Charles Leiber, decided to link it up with nanotubes that were one to a few nanometers in diameter. The idea was to replace the conventional AFM probe with nanotubes fitted with different chemical tips, such as molecules that might interact with those on the target surface. By detecting binding forces between these chemical tips and the target surface, he reasoned, one might gain new information about the surface chemistry.

Another motivation was the nanotube's tiny diameter. The team hypothesized that they could significantly increase AFM resolution by replacing its standard tip, which has a radius of curvature of 5 to 20 nanometers, with that of a carbon nanotube 0.5 nanometers in radius. "A blunt tip makes it difficult to detect fine objects. It's like trying to feel or pick up a grain of rice with a boxing glove," explains Joselevich. The study, published in *Nature*, showed that he was on the mark. The new approach successfully detects the presence of specific molecules on the target surface.

Likewise, having added a temporal resolution system that photographed the interaction every 30 seconds, Joselevich is able to view live, moving DNA as it undergoes various biochemical processes. This realtime view offers an important advantage over electron microscopy, which can be applied only to dead samples, since the high flow of electrons intrinsic to that technology destroys the sample.

Potential future applications of chemical force microscopy range from basic research aimed at a better understanding of gene expression, to the ability to detect a person's susceptibility to disease, to industrial applications, including the detection of chemical impurities on semiconductor chips.

XANES IN NANOBIOLOGY

The most sophisticated materials and machines operating on our planet are not manmade: they are produced by living organisms at the molecular, the nano-, micro- and macroscopic scales. These "devices" not only master materials synthesis and materials performance, but they also excel at quality control and self-repair. There is a lot to learn about and from biomaterials, as we are only just beginning to understand some of the fundamental mechanisms underlying the formation of biostructures, such as templation and self-assembly.

In summary, living organisms evolved to harness what we know as the laws of chemistry and physics to build the building blocks of life. They often do so with astounding efficiency: in biomineral tissues (bone, teeth, as well as the shells of crustaceans, mollusks and avian eggs, etc.), for instance, only a few per cent of the mass is organic, while the rest are accurately controlled but self-assembled minerals.

The ultimate challenge for the nano-biologist is to correlate *form, function* and *location* of molecular structures in a cell or a tissue. Methods that can explore the chemical composition, the structure, the orientation and the position of bio-components are therefore the most desirable in the nano-biologist's toolbox. X-ray PhotoElectron Emission spectroMicroscopy (X-PEEM) combined with X-ray Absorption Near-Edge Structure (XANES) spectroscopy (= spectromicroscopy) can do all this, and more: it also reveals how different molecular structures influence one another at organic-inorganic interfaces. Furthermore, spectromicroscopy can elucidate the mechanisms of formation of biominerals, thus advancing materials science with the synthesis of novel bio-inspired materials.

Here we discuss two particular recent experiments in nanobiology, in which we revealed that XANES spectroscopy is sensitive to the specific macromolecular structure of amyloid-forming molecules associated with diseases, and to the crystal orientation in mollusk shells. Interestingly, in both cases, unexpected spectroscopic results were first observed in the bio-system, then fully studied, characterized and explained in model, abiotic samples, then the spectroscopic interpretation was used to go back to the biosystem and produce biologically relevant results.

XANES DETECTS PROTEIN MISFOLDING AND AGGREGATION

We examined the ability of carbon K-edge XANES spectroscopy to detect protein and polypeptide conformation. Figure 1 shows spectra collected from poly-L-Lysine (PLL), a synthetic homopolypeptide that can be folded into a-helical or (}-pleated sheet conformations but also forms cross-P fibrils, the molecular assembly commonly called amyloid, which is common to other naturally misfolded and aggregated proteins, associated with Alzheimer's, Parkinson's, and Mad Cow diseases. In cross-P fibrils, P strands run perpendicular to the fibril axis.

These spectra are almost indistinguishable, indicating that, at least in PLL, carbon K-edge XANES spectroscopy cannot distinguish between a and P conformations. In Figure we present XANES spectra collected from a-PLL and cross-P-PLL (PLL samples kept, respectively, for 0 and 2 days at pH 11.1 and 65°C). The comparison of these two spectra demonstrates that carbon K-edge XANES spectroscopy can clearly distinguish cross-P fibrils.

The most significant difference is the enhancement of peak 2 at 287 eV, "the amyloid peak", and a shift in energy of peak 3 at –288 eV.

Peak fitting enables the quantification of spectral features and was applied to all spectra. Interestingly, the single Lys spectrum does not significantly differ from the PLL spectrum, in which 1,000-2,000 Lys residues are present.

To ensure that the spectral differences observed in Figure are due to cross-P fibril formation and not to other chemical or physical effects, we conducted extensive and systematic radiation damage studies on PLL. These revealed that while peaks 1 and 5 are radiation-damage induced, peak 2 is not.

Carbonates in particular (peak 5), accumulate due to the

harsh chemical conditions necessary to form cross-P-PLL and prolonged storage or under increasing exposure to 288.2 eV radiation, but have nothing to do with the presence or absence of fibrils. We conclude that the peak 2 enhancement and peak 3 energy shift observed in Figure cannot be radiation damage induced.

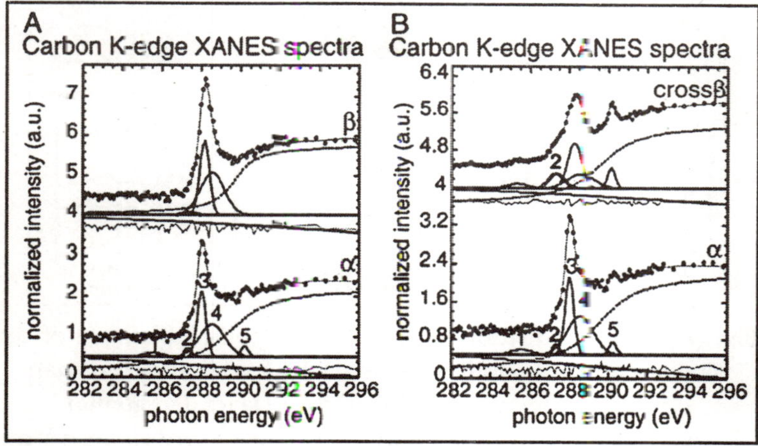

Fig. (A) Carbon K-edge XANES spectra for a-PLL (day 0 at pH 11.1) and p-PLL (day 4 at pH 11.1 and 65°C, supernatant) show that XANES spectroscopy cannot determine protein folding. (B) Carbon K-edge XANES spectra from a-PLL and cross-P-PLL shows that XANES spectroscopy detects the cross-P fibril through the enhancement of peak 2, "the amyloid peak".

After discovering the sensitivity of XANES to the cross-P structure in synthetic PLL, we extended the study to more relevant, and more complex natural proteins, which also form the cross-P structure. These include (i-amyloid peptide (A|3, residues 1-40), prion protein extracted from scrapie-infected (PrPSc) and uninfected hamster brains (PrPc), and the Sup35 yeast prion's amyloid-forming nucleus, containing only 7 amino acids: GNNQQNY.

In all systems examined, peak 2 enhancement occurred in the cross-P form of the protein. Additional radiation damage studies on A|3 confirm that peak 2 is independent of radiation dose and therefore can be confidently assigned to fibril formation.

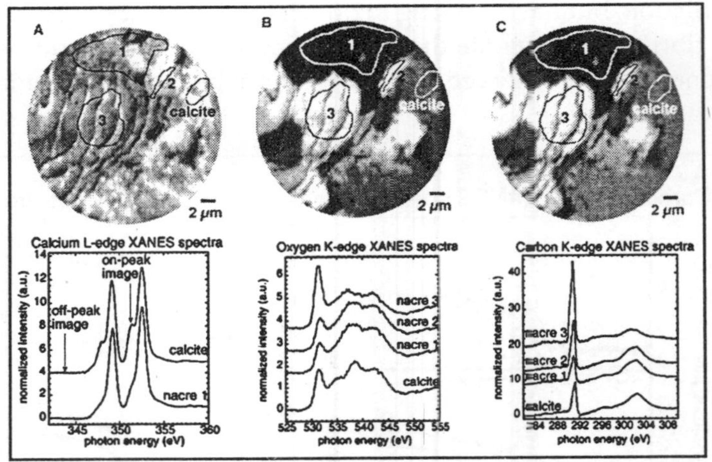

Fig. (A) Calcium distribution map of the polished surface of red abalone at the boundary between the outer shell prismati clayer (calcite) and the inner nacre layer (aragonite).

The Ca map was obtained by digital ratio of two X-PEEM images acquired with SPHINX, on- and off-peak. The on-peak was acquired at 351.3 eV, on the crystal field peak, which is strong in calcite and weak in aragonite.

Below the distribution map are the XANES spectra extracted from a representative nacre tablet (all Ca spectra from different nacre tablets are alike), and a prismatic column, showing that calcite is on the right of the border and aragonite on the left.

Distribution maps of the same region as seen in (A) for the oxygen and carbon *n* signals obtained by digital ratio of images at 531.5 eV and 516 eV, and 290.2 eV and 278 eV, respectively. Below the oxygen and carbon maps are XANES spectra extracted from the correspondingly labeled regions within the maps, showing that the tablets in the nacre region have strongly varying intensities of the 71* peak.

These results demonstrate for the first time the ability of XANES to distinguish between normally folded proteins and misfolded proteins aggregated into cross-P fibrils. XANES spectroscopy is, therefore, an additional useful method to examine protein misfolding and aggregation.

XANES Detects Orientational Disorder in Nacre Crystalline Tablets

We examined *Haliotis rufescens* (red abalone) using a combination of X-PEEM and XANES spectroscopy. In doing so, we imaged and extracted XANES spectra from individual nacre (mother-ofpearl) and prismatic tablets using the Spectromicrsocope for PHotoelectron Imaging of Nanostructures with X-rays (SPHINX).

Different grey levels the enhancement (darker grey) of the crystal field peak at 351.3 eV, which is more prominent in the calcium L-edge XANES spectrum of calcite and than in aragonite. Despite the poor energy resolution of the calcium spectra extracted from this region, it is possible to identify the imaged region as the nacre-prismatic boundary in the shell, with nacre (aragonite) on the left and the outer prismatic shell layer (calcite) on the right.

The intensity of the 7C* peak depends on the crystallographic orientation of each nacre or prismatic tablet with respect to the linearly polarized light from the synchrotron. Additionally, we found that the tablet contrast observed near the nacre-prismatic boundary continued several hundreds of microns into the nacreous region.

Immediately below Figures are XANES spectra taken at the oxygen and carbon K-edges, respectively, from correspondingly labeled tablets of Figures. These spectra quantify the 7C* peak intensity difference observed in the distribution maps and led us to investigate the polarization-dependence of XANES spectroscopy of carbonates. The contrast in Figure is due to crystallographic orientation disorder in nacre tablets. Specifically, the directions of the *c* axes of different nacre tablets differ, consequently the polar angle, that is, the angle between the *c* axis and the polarization vector varies from tablet to tablet. The observed variation could, therefore, be described as polar angle disorder. We produced theoretical XANES spectra using FEFF.

The simulated spectra exhibit a strong dependence of the 7C* peak intensity on polar orientation. To conclusively assign the source of contrast to polar rotation, however, we collected carbon K-edge XANES spectra from natural aragonite crystals. Spectra were obtained for polar rotation angles varying between 0° and 90°, and found to show dramatically different 7C* intensities.

Although a quantitative assignment of nacre tablet orientation angles is not yet possible, the results on aragonite confirm that the contrast observed in nacre is due to polar disorder. The strong crystal orientation dependence of carbon K-edge XANES spectra enabled us to map individual tablets, or stacks of tablets, in red abalone due to disorder in tablet orientation. Further investigation into this contrast-generating mechanism is necessary for quantitative results. For the moment, its discovery led to new insights into the nacre formation mechanism, which has been and remains elusive.

Capter 2

Nanoorganisms

MICROBIOLOGY

Microbiology is the study of microorganisms, that is the organisms which are of microscopic dimensions. These organisms are too small to be clearly perceived by the unaided human eye. If an object has a diameter of less than 0.1 mm, the eye can not perceive it at all and very little detail can be perceived in an object with a diameter of I mm. Roughly speaking organisms with a diameter of 1 mm or less are microorganisms and fall into the broad domain of microbiology. Since most microorganisms are only a few thousandths of a millimeter in size. they can only be seen with the aid of microscope.

As a direct consequence of the invisibility of micros to the naked eye and the need for special techniques to study them, microbiology was the last of the three major divisions in biology the other two are botany and zoology to develop. Microbiology is the study of organisms that are too small to be clearly seen by the unaided eye. Roughly speaking, organisms with a diameter of 1 mm or less which can not be seen by the human eye unaided, are called microorganisms. These include protozoa, algae, fungi and bacteria Viruses are ultramicroscopic and have an obligate parasitic relationship, but for practical purposes these still come under the domain of microbiology.

At present, there is general agreement to include five major groups as microorganisms: the subdivisions of virology, bacteriology, mycology, phycology and protozoology

Traditionally, all organisms had been included under the disciplines of either botany or zoology. Bacteria, algae and fungi have in the past been considered to be the part of the Plant Kingdom while protozoa have been included in the Animal Kingdom. This view cannot be supported from a taxonomic standpoint. Since microbiology encompasses the study of groups of organisms in all three of these divisions of biology it may be argued that it covers a greater biological diver itv than the other two divisions.

Although microorganisms have existed for a long time, their existence was unknown until the invention of the microscope in the 17th century. The year 1674, marks the birth of microbiology when Antony van Leeu-wenhoek, a Dutch cloth merchant, looked at a drop of lake water through a glass lens which he had ground.

What he observed through this simple magnifying lens was an amazing sight since that was perhaps the first time that man ever had a glimpse of the world of the microbes. Subsequently, in a series of letters to the Royal Society of London, Leeuwenhoek describ-ed a variety of microorganisms such as protozoa, algae, yeast and bacteria and the description was so precise that it is now possible to assign them into specific genera without any additicnal descrip-tion.

Leeuwenhoek had little formal education but his keen interest in nature made him to examine a variety of materials. Glass grinding and preparation of lenses was his hobby and this led him to the assembly of about 400 simple microscopes. The earliest microscope that he constructed consisted of a spherical lens mounted on two plates.

MICROBES AND CHEMICAL CHANGES

At about this time Liebig the German chemist, had claimed that fermentation and putrefaction were processes caused by the chemical instability of organic molecules. On the other hand, Pasteur had shown that living yeast cells were required for the conversion of sugar into alcohol and had suggested that these fermentations and transformations are brought about by the agency of microorganisms.

At this time. the distilleries in France, where alcohol was produced from sugar beet had difficulties in their fermentations and requested Pasteur for help. Pasteur found that the trouble was due to the replacement of the alcoholic fermentation by a lactic fermentation. Instead of finding yeast in the fermenting brew, he found rod-shaped structures.

He concluded that these rod shaped organisms were responsible for the conversion of sugar into lactic acid. Subsequently, he also showed that different fermentations are brought about by specific microorganisms. For example, the alcoholic fermentation was the result of living yeast cells while the lactic acid fermentation was the result of the activity of lactic bacteria.

Interestingly in 1897, after Pasteur, Buchner showed that intact yeast cells were not. required for the conversion of sugar into alcohol and carbon dioxide He prepared a cell free yeast juice by grinding yeast cells with sand and showed that this cell free juice had the ability to carry out the alcoholic fermentation. He concluded that the active agent in the yeast cells responsible for bringing about the fermentation was a protein and he named it as zymase. This finding gave support to Liebig's thesis that living cells were not required to bring about fermentations and marks the beginning of the study of enzyme chemistry.

ANCIENT MICROBIOLOGICAL HISTORY

Ancient man recognized many of the factors involved in disease. Early civilizations on crete, india, pakistan and scotland invented toilets and sewers; lavatories, dating around 2800 bc, have been found on the orkney islands and in homes in pakistan about the same time. One archaeologist has stated that "the high quality of the sanitary arrangements could well be envied in many parts of the world today". In rome, 315 ad, the public lavatories were places where people routinely socialized and conducted business.

Ten to twenty people could be seated around a room, with their wastes being washed away by flowing water; it must have been difficult to "*stand on your dignity*" under such circumstances. The chinese used toilet paper as early as ad 589. In europe moss, hay and straw were used for the same purpose. I can personally attest to the use as late as 1962 of "slick magazines" as toilet paper in certain european camp grounds. The first cities to use water pipes (of clay) were in the indus valley of pakistan around 2700 bc. Metal water pipes were used in egypt (2450 bc) and the palace of knossos on crete around 2000 bc had clay pipes.

Rome built elaborate aqueducts and public fountains throughout its empire to insure a clean supply of water for its citizens. Rome had a "water commissioner" who was responsible for seeing that the water supply was kept adequate and clean; the punishment for pollution of the water supply was death. Lead was commonly used for roman pipes and the subsequent fall of the roman empire has been related by some to the effects of lead on the roman brain.

Most ancient peoples recognized that some diseases were communicable and isolated individuals thought to carry "infections". An example of this is the universal shunning of lepers, which occurs even today. When the black death struck europe, entire villages were abandoned as people fled in an effort to escape the highly infectious plague.

Similarly, in the middle ages the rich of europe fled to their country homes when small pox struck in an effort to escape its terrible consequences. The fact that people who recovered from a particular disease were immune to that disease was probably recognized many different times in many places. Often these survivors were expected to nurse the ill. Greek and roman physicians routinely prescribed diet and exercise as a treatment for ills.

Sadly, we know that this knowledge did not help most of our ancestors and that the human life span was, until the last 200 years, more often than not cut short due to infectious disease. Even today approximately 15,000,000 children die per year, mainly from infectious diseases that are preventable with basic sanitation, immunization and simple medical treatments. One might honestly question just how far we have come in our treatment of disease. An excellent synopsis of the history of microbiology ancient people had certainly seen masses of microbes, such as mold and bacterial colonies, on spoiled food, but it is doubtful if anyone considered that they were viewing living organisms.

Small boys and maybe a few love-sick adults staring into a clear pond, must have seen tiny specks moving rapidly about and some may have considered them living creatures, but to express this to their friends would be equivalent to us telling our friends that we'd seen a flying saucer.

The first person to report seeing microbes under the microscope was an englishman, robert hooke. Working with a crude compound microscope he saw the cellular structure of plants around 1665. He also saw fungi which he drew. However, because his lens were of poor quality he was apparently unable to "see" bacteria.

Anton van leeuwenhoek was a man born before his time. Although not the first to discover the microscope or to use magnifying lens, he was the first to see and describe bacteria. We know that he was a "cloth merchant" living in delft holland. And that he used magnifying lens to view the quality of the weave of the merchandise he purchased. He traveled to england in 1668 to view english cloth and there he saw drawings of magnifications of cloth much greater than any of the current lens available in holland would do.

He returned to holland and took up lens grinding. Being meticulous, he developed his lens grinding to an art and in the process tested them by seeing how much detail he could observe with a given lens. One can guess that he chanced to look at a sample of pond water or other source rich in microbes and was amazed to see distinct, uniquely shaped organisms going, apparently purposefully, about their lives in a tiny microcosm. He made numerous microscopes from silver and gold and viewed everything he could including the scum on his teeth and his semen.

His best lens could magnify ~300-500 fold which allowed him to see microscopic algae and protozoa and larger bacteria. He clearly had excellent eyesight because he accurately drew pictures of microbes that were at the limit of the magnification of his lens. He used only single lens and not the compound lens of the true microscopes we employ today; which makes his observations all the more amazing. He wrote of his observations to the royal society of london in 1676 and included numerous drawings.

He astonished everyone by claiming that many of the tiny things he saw with his lens were alive because he saw them swimming purposefully about. This caused no end of shock and wonderment and numerous people hurried to delft to see if this dutchman was "in his cups" or if he was really onto something new and wonderful.

A few minutes with one of his numerous microscopes was all it took to convert his visitors to enthusiastic believers in the existence of these tiny beasties living all around them. His discovery was the equivalent of our finding life on mars today.

THE CLEAN NUT AND THE REVENGE OF THE BACTERIA

In the 1800s people began to use hospitals. Hospitals also became centers of physician training. In 1841 young doctor ignaz semmelweis was hired to run a maternity ward in a vienna hospital. There were two birthing wards in his preview, one run by midwives and the other by doctors. Semmelweiss noticed that the death rate among mothers in the doctor's ward ran as high as 18% from the blood infection known as child bed fever or puerperal sepsis, whereas in the midwife ward the death rate was much lower.

When he suggested that the doctors might contribute to this he was fired. Subsequently rehired, he saw a friend die from puerperal sepsis after cutting himself during an autopsy of a patient who had died of puerperal sepsis. He reasoned that there was an invisible agent that caused both deaths and that one could transfer it from the autopsy room to the birthing rooms and thus infect the mothers during birthing.

Acting on this assumption, semmelweis instituted sanitary measures which included having the doctors wash their hands in disinfectant and change from lab coats dripping with pus and blood from the autopsy room to clean lab coats before examining patients or assisting in a birth. The death rate of the mothers dropped by 2/3 in his ward. However, the other doctors objected so strongly to his rules that semmelweis was again fired and left vienna.

He took other hospital jobs where he instituted the same standards of cleanliness which resulted in the same decline in deaths and a revolt of his fellow physicians. Semmelweis ended up dying in an insane asylum from a blood infect that resembled puerperal sepsis.

SPONTANEOUS GENERATION

The mystery of life has puzzled and confounded humans since the first human began to contemplate his world. The religions of ancient societies were built around the seasons, the sun and the renewal of life as these were so clearly tied to survival; both of the human species through birth and death and of the individual in the attainment of sustenance. Spontaneous generation or the idea that life routinely arises from non-life was a common sense explanation of the miracle of life.

It had the advantage of simplicity, ease of understanding and didn't require any waste-of-time thinking. As science and the scientific method grew with the slow accumulation of knowledge, observant individuals began to consider the origin of life more deeply. Simple observations convinced many people that all the larger animals and plants produced life from previous life. Despite this, the mass of humans clung to the comfortable idea of spontaneous generation further, religions saw it as a convenient way to demonstrate the hand of god operating continuously in the world.

Some individuals such as j.b. Van helment even described how one could make mice from grain, a jar and dirty rags by putting them together in a dark place for a few weeks and soon mice would appear in the jar.

Other, more perceptive individuals, like f. Redi tested the common idea that maggots arise via spontaneous generation on rotting meat. He placed a piece of meat in three jars, one he left open, one he corked tightly and the third he covered with a fine mesh gauze. Maggots only appeared in the open container, no matter how long he left the jars. Redi's experiment was important because of its eloquent simplicity. Anyone could repeat it and obtain the same clear results. Nevertheless, many people clung fiercely to the idea of spontaneous generation, while others designed experiments to test it.

In every case the results of the majority of these experiments indicated that spontaneous generation did not occur. The intellectual ferment this controversy stirred up gradually evolved into the scientific method as the various antagonists questioned each other's assumptions and, more importantly, their experimental design.

These arguments forced the designing of better experiments and eventually persuaded all but the most recalcitrant believers to discard spontaneous generation as an explanation for all higher life forms. Then, in the 1800's the refinement of the microscope, through which people could see tiny life forms that they assumed were simple, gave spontaneous generation proponents new life.

Again, flawed common sense led reasonable people astray. As the existence of microscopic life was accepted, the assumption was that such life must be simple compared to higher, more complex life. The reasoning that followed this erroneous assumption was that since the microbes were small they must be simple & it followed that they were formed by spontaneous generation, hence god was still at work creating micro-life. Small is not simple!

The battle over spontaneous generation raged anew both from the pulpit and the lab. A number of scientists performed elementary experiments in which they treated soups and broth's, which left unheated would team with microbes after a few days, with heat to destroy any life present in them and asked the question: "Would new life arise in these sterile soups"? Spallanzani boiled "Soup" in glass containers and melted the glass closed.

The observation that nothing subsequently grew in this

"heated" soup suggested that spontaneous generation didn't work. His detractors, rightly criticized his experiments, proposing that since air is necessary for life and since he had sealed the flask to air, obviously no life could develop. Others boiled soups and microbes grew, thus apparently supporting spontaneous generation. But again the preponderance of data suggested that spontaneous generation did not even apply to the "simple" microbes.

In 1859 one of the fathers of modern microbiology, l. Pasteur decided to settle the question of spontaneous generation once and for all. A genius at devising definitive experiments, pasteur first drew the necks of glass flasks out so that they remained open to the air, but were bent so that air could only enter by a curved path. He then added broth and boiled it to destroy contaminating microbes. These flasks were then incubated and observed for months. He reasoned that the microbes in the air that could contaminate the sterile broth would be trapped on the sides of the thin glass necks before they reached the sterile broth. If spontaneous generation didn't occur no growth should take place.

This is exactly what happened, the flasks remained sterile indefinitely, until pasteur tipped the sterile broth up into the curved neck where he predicted the airborne organisms would have settled.

After doing this the broths always grew microbes. These experiments ended the spontaneous generation controversy because these experiment was so elegant and simple and the results so clear, that anyone could repeat them.

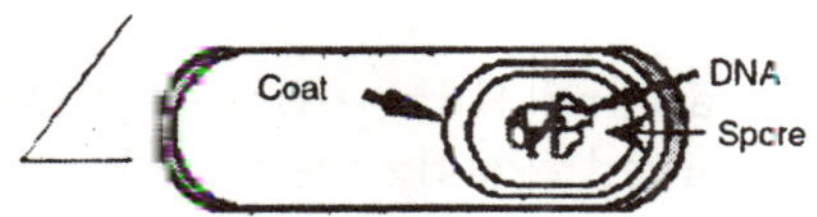

Fig. Spore Structure

Later, an earlier problem, in which occasional heated-broths did not remain sterile, was explained with the discovery of the heat resistant bacterial some of which could can survive several hours of boiling without being killed.

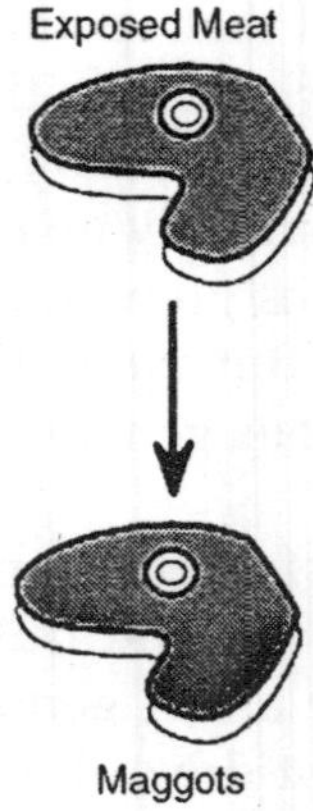

Fig. The Redi Experiment

Pasteur discovered many of the basic principles of microbiology and, along with R. Koch, laid the foundation for the science of microbiology. In 1857 Napoleon III was having trouble with his sailors mutinying because their wine was spoiling after only a few weeks at sea. Naturally Napoleon was distraught because his hopes for world conquest were being scuttled (pardon the pun) over a little spoiled wine, so he begged Pasteur for help.

Pasteur, armed with his trusty microscope, accepted the challenge and soon recognized that by looking at the spoiled wines he could distinguish between the contaminants that caused the spoilage and even predict the taste of the wine solely from his microscopic observations. He then reasoned that if one were to heat the wine to a point where its flavour was unaffected, but the harmful microbes were killed it wouldn't spoil.

As we are aware this process, today known as pasteurization, worked exactly the way he predicted and is the foundation of the modern treatment of bottled liquids to prevent their. It is important to realise that pasteurization is not the same as sterilization. Pasteurization only kills organisms that may spoil the product, but it allows many microbes to survive, whereas sterilization kills all the living organisms in the treated material.

Pasteur also realized that the yeast that was present in all the produced the alcohol in wine. When he announced this, a number of famous scientists were enraged, because the current theory of wine production was that wine formation was the result of spontaneous chemical changes that occurred in the grape juice. Pasteur was attacked furiously at scientific meetings, to the point where certain scientists did humorous skits about Pasteur and his tiny little yeast "stills" turning out alcohol. Pasteur had the last laugh however as people all over the world soon realized that if he was right they could control the quality of wine by controlling the yeast that made it. In a short period many others verified his observations and the opposition sank without a sound.

Pasteur discovered many of the basic principles of microbiology and, along with R. Koch, laid the foundation for the science of microbiology. In 1857 Napoleon III was having trouble with his sailors mutinying because their wine was spoiling after only a few weeks at sea. Naturally Napoleon was distraught because his hopes for world conquest were being scuttled (pardon the pun) over a little spoiled wine, so he begged Pasteur for help.

Pasteur, armed with his trusty microscope, accepted the challenge and soon recognized that by looking at the spoiled wines he could distinguish between the contaminants that caused the spoilage and even predict the taste of the wine solely from his microscopic observations. He then reasoned that if one were to heat the wine to a point where its flavour was unaffected, but the harmful microbes were killed it wouldn't spoil.

As we are aware this process, today known as pasteurization, worked exactly the way he predicted and is the foundation of the modern treatment of bottled liquids to prevent their. It is important to realise that pasteurization is not the same as sterilization. Pasteurization only kills organisms that may spoil the product, but it allows many microbes to survive, whereas sterilization kills all the living organisms in the treated material.

Pasteur also realized that the yeast that was present in

all the produced the alcohol in wine. When he announced this, a number of famous scientists were enraged, because the current theory of wine production was that wine formation was the result of spontaneous chemical changes that occurred in the grape juice. Pasteur was attacked furiously at scientific meetings, to the point where certain scientists did humorous skits about Pasteur and his tiny little yeast "stills" turning out alcohol. Pasteur had the last laugh however as people all over the world soon realized that if he was right they could control the quality of wine by controlling the yeast that made it. In a short period many others verified his observations and the opposition sank without a sound.

MICROBES

A microbe is any living thing that spends its life at a size visible sometimes only with a microscope. It is too tiny to be seen with the naked eye. Microbes are the oldest form of life on Earth. Some types have existed for billions of years. They may live as individuals or cluster together in communities. Microbes live in the water we drink, the food we eat and the air we breathe.

Right now, billions of microbes are swimming in our belly and mouth and crawling on our skin! Don't worry, over 95% of microbes are good for you. Microbes include bacteria, viruses, fungi, algae and protozoa. These single-cell organisms are invisible to the eye, but they can be seen with microscopes.

Fig. Microbe

The term microbe is short for microorganism, which

means small organism. To help people understand the different types of microbes, they are grouped or classified in various ways. Microbes are very diverse and represent all the great kingdoms of life. In fact, in terms of numbers, most of the diversity of life on Earth is represented by microbes.

CLASSIFICATION OF MICROBES

The term microbe is short for microorganism, which means small organism. To help people understand the different types of microbes, they are grouped or classified in various ways. Microbes are very diverse and represent all the great kingdoms of life. In fact, in terms of numbers, most of the diversity of life on Earth is represented by microbes.

Here is an outline of the major groups of microorganisms:

- Viruses
- Bacteria
- Algae
- Fungi
- Protozoa

Viruses

A virus is too small to be seen without a microscope. A virus is basically a tiny bundle of genetic material carried in a shell called the viral coat. Some viruses have an additional layer around this coat called an envelope. That's basically all there is to viruses.

There are thousands of different viruses that come in many shapes. Many are multi-sided or polyhedral. If you've ever looked closely at a cut gem, like the diamond in an engagement ring, you've seen an example of a polyhedral shape.

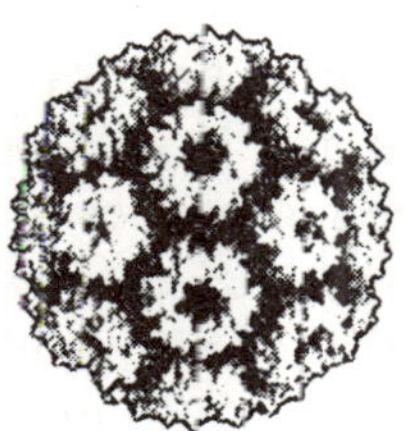

Fig. Viruses

Unlike the diamond in a ring, however, a virus does not taper to a point, but is shaped the same all around. Other viruses are shaped like spiky ovals or bricks with rounded corners. Some are like skinny sticks while others look like pieces of looped string. Some are more complex and shaped like little spaceship landing pods.

Viruses are found on or in just about every material and environment on Earth from soil to water to air. They're basically found anywhere there are cells to infect. Viruses can infect every living thing. However, viruses tend to be somewhat picky about what type of cells they infect. Plant viruses are not equipped to infect animal cells, for example, though a certain plant virus could infect a number of related plants. Sometimes, a virus may infect one animal and do no harm, but cause a great deal of damage when it gets into a different but closely related animal.

Viruses exist to reproduce only. To do that, they have to take over suitable host cells. Upon landing on a suitable host cell, a virus gets its genes inside the cell either by tricking the host cell to pull it inside, or by connecting its viral coat with the host cell wall or membrane and releasing its genes inside, or by injecting their genes into the host cell's DNA. The viral genes are then copied many times, using the process the host cell would normally use to reproduce its own DNA. The new viral genes then come together and assemble into whole new viruses. The new viruses are either released from the host cell without destroying the cell or eventually build up to a large enough number that they burst the host cell.

Bacteria

Bacteria consist of only one cell, but they're a very complex group of living things. Some bacteria can live in temperatures above the boiling point and in cold below the freezing point.

There are thousands of species of bacteria, but all of them are basically one of three different shapes. Some are rod- or stick-shaped; others are shaped like little balls. Others still are helical or spiral in shape. Some bacteria cells exist as individuals while others cluster together to form pairs, chains, squares or other groupings. Some bacteria can make their own food from sunlight, just like plants. Also like plants, they give off oxygen. Other bacteria absorb food from the material they live on or in. Some of these bacteria can live off iron or sulfur! The bacteria that live in our stomach absorb nutrients from the digested food you've eaten.

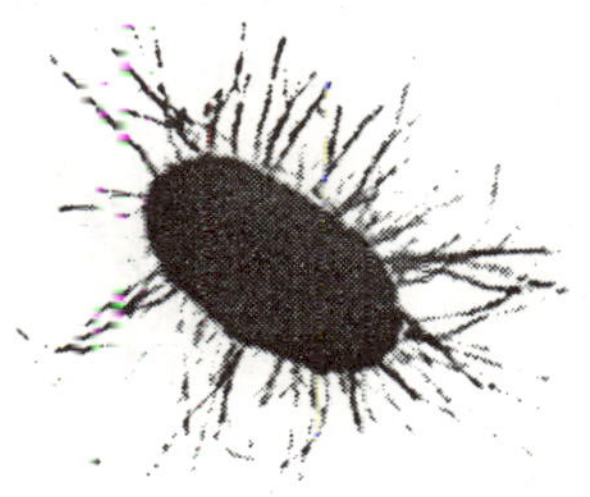

Fig. Bacteria

Some bacteria move about their environment by means of long, whip-like structures called flagella. They rotate their flagella like tiny outboard motors to propel themselves through liquid environments. They may also reverse the direction in which their flagella rotate so that they tumble about in one place.

Other bacteria secrete a slime layer and ooze over surfaces like slugs. Others stay almost in the same spot. Bacteria live on or in just about every material and environment on Earth from soil to water to air and from our body to the Arctic ice to the Sahara deserts. Each square centimeter of our skin averages about 100,000 bacteria. A single teaspoon of soil contains more than a billion (1,000,000,000) bacteria.

Algae

Algae are found in fresh and salt water around the world. They can also grow on rocks and trees and in soil when enough water is available.

Most algae are able to make energy from sunlight, like plants do. They produce a large amount of the oxygen we breathe. However, at some stages of their lives, some algae get their nutrients from other living things.

Diatoms are one kind of algae. They have hard shells made out of glass. When they die, these shells sink to the bottom of their watery environments.

We mine deposits of these glass shells that formed hundreds of thousands of years ago to make abrasives, shiny road paint and grit in toothpaste. Diatoms come in all sorts of shapes—some are round and others are oval. Some look like leaves and others like fat commas

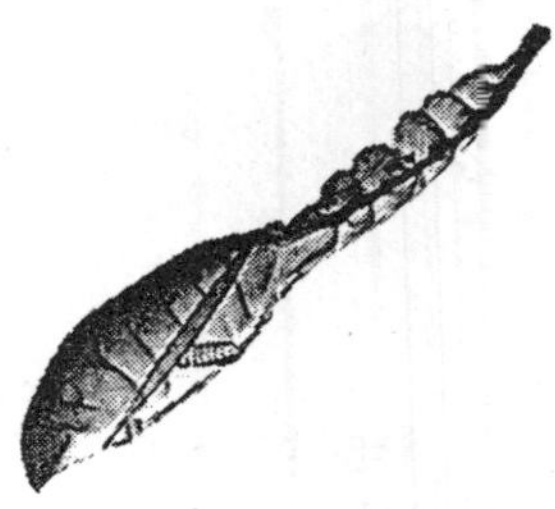

Fig. Algae

Because algae make so much oxygen, these microbes are very helpful. But sometimes certain kinds of algae can also grow in such large numbers that when they suddenly die off en masse, the breaking down of their cells by bacteria destroys the amount of dissolved oxygen in the water, hurting the animals and plants that live there.

Fungi

Fungi are organisms that scientists once confused with plants. However, scientists have found that, at the cell level, the fungi are more like animals than they are like plants. For one thing, fungi cannot synthesize their own food like plants do, but instead they eat other organisms as do animals. Fungi come in a variety of shapes and sizes and different types. They can range from single cells to enormous chains of cells that can stretch for miles.

Fungi include single-celled living things that exist individually, such as yeast and multicellular clusters, such as molds or mushrooms. Yeast cells look round or oval under a microscope. They're too small to see as individuals, but we can see large clusters of them as a white powdery coating on fruits and leaves.

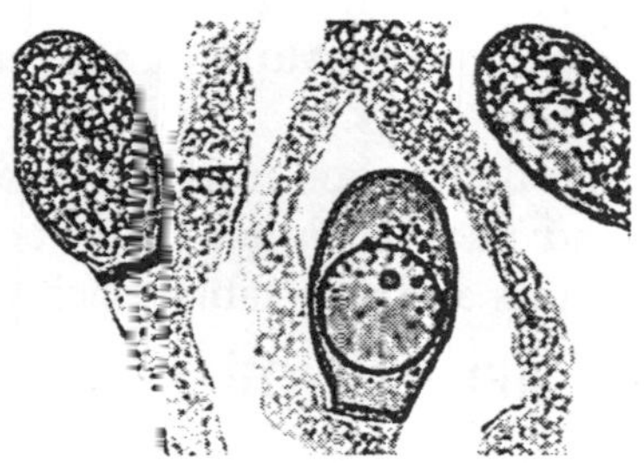

Fig. Fungi

Molds are described as filament-like because they form long filament-like, or thread-like, strands of cells called hyphae. These hyphae are what give mold colonies their fuzzy appearance. They also form the fleshy body, or mushroom, that some species grow. It may seem strange to think of something as big as a mushroom as a microbe. But the cells of the hyphae making up that mushroom are connected in a closer way than the cells of other multicellular living things, like we and me, are.

Fungi usually grow best in environments that are slightly acidic. They can grow on substances with very low moisture. Fungi live in the soil and on our body, in our house and on plants and animals, in freshwater and seawater. A single teaspoon of topsoil contains about 120,000 fungi.

Fungi are basically stationary. But they can spread either by forming reproductive spores that are carried on wind and rain or by growing and extending their hyphae. Hyphae grow as new cells form at the tips, creating even longer chains of cells. Fungi absorb nutrients from living or dead organic matter that they grow on. They absorb simple, easily dissolved nutrients, such as sugars, through their cell walls. They give off special digestive enzymes to break down complex nutrients into simpler forms that they can absorb.

Some fungi are quite useful to us. We've used several kinds to make antibiotics to fight bacterial infections. These antibiotics are based on natural compounds the fungi produce to compete against bacteria for nutrients and space. We use baker's yeast, to make bread rise and to brew beer. Fungi break down dead plants and animals and keep the world tidier. We're exploring ways to use natural fungal enemies of insect pests to get rid of these bugs.

There are some dangerous fungi that cause diseases in plants, animals and people. Fungi ruin about a quarter to half of harvested fruits and vegetables each year.

MICROBES ARE HELPFUL OR NOT

Microbes are much more our friends than our enemies. Although some microbes cause health problems such as strep throat, chickenpox and the common cold, most microbes make our lives better such as: Bacillus thuringiensis - a common soil bacterium that is a natural pest-killer in gardens and on crops. Arbuscular mycorrhizas - fungus living in the soil that helps crops take up nutrients from the soil. Accharomyces cerevisiae - baker's yeast that makes bread rise.

Escherichia coli - one of many kinds of microbes that live in our digestive system to help we digest our food every day. Streptomyces - bacteria in soil that makes an antibiotic used to treat infections. Pseudomonas putida - one of many microbes that clean wastes from sewage water at water treatment plants. Lactobacillus acidophilus - one of the bacteria that turn milk into yogurt.

There are many other important jobs microbes do. They are used to make medicine. They break down the oil from oil spills. They make about half of the oxygen we breathe. They are the foundation of the food chain that feeds all living things on earth. We've been using microbes for thousands of years to make products we need and enjoy. For example, we can thank fungi for the cheese on our cheeseburger and yeast for our bun. Cheese and bread are two microbe-made foods people have been enjoying since time began.

In pollution control, researchers are using bacteria that eat methane gas to clean up hazardous waste dumps and landfills. These methane-eating bacteria make an enzyme that can break down more than 250 pollutants into harmless cells.

By piping methane into the soil, researchers can increase growth of the bacteria that normally live in the polluted soil. More bacteria means faster pollution break up. Also, bacteria is being used as one of the tools to clean up oil spills. These bacteria eat the oil, turning it into carbon dioxide and other harmless by-products. Fungi and bacteria produce antibiotics such as penicillin and tetracycline.

These are medicines we use to fight off harmful bacteria that cause sore throats, ear infections, diarrhea and other discomforts. Scientists have changed the genetic material of bacteria and yeasts to turn them into medicine. They inject genes for medicines they want to make into the microbe cells, as if adding new building information to the microbe's cell DNA. The scientists then grow the microbes in huge containers called fermenters where they reproduce into billions, all making new medicines.

Microbes and the Days of Creation

The world of germs and microbes has received much attention in recent years—and for good reason. We frequently hear the term microbe associated with organisms such as Escherichia coli, Salmonella, anthrax bacteria, antibiotic-resistant tuberculosis, MRSA, HIV, malaria, Stachybotrys and other microscopic creatures. But where do microbes fit into the creation account? Were they created along with the rest of the plants and animals in the first week of creation, or were they created later, after the Fall?

Are microbes a result of the Curse? These and other questions are some that a group of professional creation microbiologists have been asking and their answers may surprise us. Ongoing research based on the creation paradigm appears to provide some answers to these puzzling questions.

Although we cannot be dogmatic about the details of microbe origin during Creation Week, it is believed that a reasonable extrapolation from biological data and Scripture can be made about the nature of microbes in a fully mature creation. Past creation scientists such as Leeuwenhoek, Pasteur and Lister, were blessed by God as He revealed to them critical insight into His creation. So, where do these microbes fit into the very good days of creation? Before answering this question, three terms, microbe, germ and symbiosis, need to be defined. These are relatively new or "modern" terms. First, the Bible does not use these specific terms. These terms were not commonly used until the end of the nineteenth century.

The term microbe was first used in 1878 to describe "extremely minute living beings." Before 1878, scientists including Louis Pasteur, used a variety of terms rather loosely to label the very small organisms that had interested them. It was not clear whether microbes belonged to the animal or plant kingdom, or to a completely different one. The term microbe was given by Charles E. Sedillot to describe bacteria Later, it would also be used of eukaryotic cells, including algae, fungi, protozoans and slime molds. Some people refer to viruses as microbes, but others do not because viruses are not cells. Viruses have nonliving, as well as living characteristics.

Today, the term germ refers to disease-causing microbes, or pathogens. All germs would have originated after the Fall The Edenic Curse would have profoundly influenced all creation, including viruses, bacteria, fungi and protozoans that would later become pathogens or parasites. The origin of infectious disease is complex and multifaceted. This topic is further explored in the book. The Genesis of Germs It provides some understanding into the origin of infectious diseases. From a biblical worldview, infectious diseases and patho-genesis are a secondary state in nature. It is not the way the Creator intended for man and nature. Most microbes are beneficial to man and nature. Only about 5–10% of all bacteria are pathogenic.

Many microbes live in a mutualistic relationship with plants, animals, humans and other microbes. Mutualism is a type of symbiosis. The term symbiosis is used to describe an intimate association between organisms of different species. Some symbiotic relationships between microbes and plants, animals and humans are essential for life on earth.

For this reason most microbiologists maintain that bacteria, fungi, protists and other microbes have been maligned in the news media. Without our intestinal flora, we would not digest food nor acquire vitamins and minerals very efficiently. Without fungi, bacteria, algae and protozoans, life on earth could not last. This is because microbes provide essential "services" (e.g., nitrogen fixation, nitrification and denitrification) in nutrient cycles.

Viruses Fit into Creation

The determination of virus origin is uncertain. It may be that viruses have multiple origins. Some may be degenerate parts from cells after the Curse; still others may have their origin during the days of creation. Today, we think of viruses (Latin for "poison") only in the context of disease. However, some viruses (or at least virus-like genes) are involved in a positive function in nature.

Some groups of viruses, like bacteriophages, play a positive role in controlling bacteria in ecosystems and may play a role in diversity. Another group of viruses play a role in turning off the immune system during pregnancy in mammals and humans. This is a group referred to as endogenous retroviruses (ERVs). ERVs are among a kind of repetitious genetic elements called "retrotransposons". Research has shown that the ERV design prohibits the mother's immune system from damaging the child's body.

These retroviruses cannot fully replicate, only expressed in local immune cells (such as macrophages) of the placenta, thereby preventing them from initiating a full-blown immune response. Thus, the mother's immune system remains competent to respond to other infections but is specifically prevented from mounting an immune response to the developing embryo.

So in creation, the selective ability to turn off the immune system for protection would be a "good" design. Other ERVs also play a positive role in animal and human reproduction. However, since the corruption of creation, the corrupted retrovirus, HIV and various leukemia viruses turn off the entire immune system, leaving the body open to devastating infections.

These examples may provide clues to the origin of viruses and how some may have been created during Creation Week by design and how some have been corrupted as a result of the Fall.

Aseptic Technique

Trying to study this mixed population is often difficult and in the tradition of the scientific method; researchers dissect a system and study each piece in isolation. For microorganisms this means separating the organisms and getting them into pure culture. A pure culture is defined as a growth of microorganisms (a culture) that contains one cell type. It is essential in microbiology to be able to obtain and preserve pure cultures.

Over 100 years ago, Robert Koch devised methods to achieve this goal and the methods he developed are essentially still used today. The goals of aseptic technique are two-fold. The first objective is to obtain pure cultures and secondly to prevent cross-contamination. Microorganisms in culture must not escape into the environment and microbes in the environment must not get into the cultures we are studying. It is essential that aseptic technique be understood and practiced correctly.

Contaminated cultures are worthless for diagnosis or for doing research on, because it is unclear what microbe is performing any action that is being observed. Aseptic methods commonly used are flame sterilization, tube transfer, streak plates, spread plates and pour plates. Flame sterilization is an easy method to insure sterile transfer of a culture from a source to a growth medium.

Tube transfer is useful for moving inocula from one tube to another. Mechanical dilution by making streak plates is the preferred method for obtaining a pure culture of a microorganism. Finally, spread plates and pour plates are common methods for enumerating microorganisms and are sometimes useful for obtaining isolated colonies.

Flame Sterilization and Tube Transfer

Flame sterilization is a very quick simple method of killing microorganisms on an inoculating loop or needle. The loop or needle is held inside a flame for a few seconds to bring it to redness and then cooled. Once cool, the loop or needle can be used for various culture manipulations. Make sure that the area that contacts the culture is flamed to redness. Also, be patient and let the loop cool down, this usually takes about 15-30 seconds. Learning this technique is essential to everything else we do in microbiology.

Transfer of culture from agar plates to tubes, or from tube to tube, is a common, simple procedure. It is important to perform these transfers in a consistent and rapid manner. The following protocols have been found effective. To transfer a culture from an agar plate to a broth or agar slant:

- Place the Bunsen burner in front of we and assemble all necessary equipment with in arms reach. Position everything so that we will not burn ourself while trying to inoculate our tubes.
- Label the tube of broth or agar to be inoculated with identifying marks. The culture, the date and our initials for example. Place it in a rack in front of we .
- Holding the inoculating loop handle, flame the entire wire to redness.
- When the wire cools (about 15-30 seconds) remove the lid of the plate with our other hand and obtain an inoculum by removing a small portion of the surface growth on the agar plate. In most cases we will be picking an isolated colony. Choose a well isolated one. Do not dig into the agar. Replace the lid of the plate immediately.
- Hold the tube to be inoculated with the free hand. Remove the cotton plug or cap of the tube with the little finger of the hand holding the needle holder. If a cotton-plugged tube is used, the mouth of the tube should be passed briefly through the flame to singe off dust and lint particles. (Dust or lint may fall into the tube and contaminate the medium.)

- Introduce the inoculum into the tube. When inoculating a tube of broth, rub the wire against the glass just above the fluid level and then tip the tube slightly to wash the inoculum into the broth. The wire should not be rattled against the sides of the tube to shake an inoculum into the broth; this is unnecessary and may create a dangerous aerosol. If the transfer is made to an agar slant, a single mid-line stroke over its surface is made with the wire or loop.
- Replace the cap or plug.
- Flame the inoculating wire again to redness, slowly to avoid spattering. Put the loop holder down after the wire cools.

Making a Medium

Making medium is as simple as cooking and a crude medium can be made in almost any kitchen with a few utensils and a source of heat. Below is described the production of a chicken broth medium that will grow many common microorganisms.

The medium will be sterilized by tyndaliza-tion. Simply boiling a medium once, while it will kill most vegetative cells, does not kill endospores. However, heating encourages the endospores to germinate and a second boiling kills these microbes. The boiling process is repeated a third time to ensure that all spores have germinated and been killed.

- Add 250 ml (about 1 cup) of water into a glass container or some other vessel that can stand boiling water. The container should be something we can cover. Glass bottles that can stand boiling or canning jars work well.
- To this add 15 grams, about 1 tablespoon of instant chicken broth crystals. and stir until dissolved. Cover loosely so that steam can escape, but dust and dirt cannot enter.
- Microwave the broth on high for 3 minutes or until it just begins to boil.
- Let the broth cool and sit for at least 2 hours up to

overnight at room temperature. Make sure that the medium vessel is covered to prevent contamination from the air. At this point the vegetative cells are dead and most endospores will germinate.

- Repeat steps 3 and 4 a second time. This will kill the endospores that germinated on day 1.
- Repeat steps 3 and 4 a third time. This will eliminate the remaining endospores, making the medium sterile.
- Place the medium in a warm place overnight. If our tyndalization was done correctly, our medium should remain clear and free of microbes.
- While we are waiting for our medium, search for a sample we would like to inoculate into it. Almost any sample should work, but natural samples are a good idea for this experiment. Samples coming from food or from our body may contain pathogens. Something we probably want to avoid.
- Check our medium. If it is still clear and sterile, add a pinch of our collected sample to the medium. (The exact amount really does not matter). Again place the medium in a warm place and check it periodically. After a few days, we should start to see the medium become more turbid.
- If we have the equipment, examine a sample of the medium under a microscope. What shapes of microbes are present? Is there more than one species?
- When finished, carefully dispose of the borth. It is basically a spoiled food, but the properties of the spoilage microbe are unknown and should be treated with care.

STREAK PLATES

The streak plate method is a rapid and simple technique of mechanically diluting a relatively large concentration of microorganisms to a small, scattered population of cells. The goal is to obtain isolated colonies on a large part of the agar surface, so that desired species can then be brought into pure culture. Proper streaking of plates is an indispensable tool in microbiology. In most cases a closed inoculating loop is used for streaking plates.

The wire loop should not be badly oxidized or pitted or it will fail to dilute the inoculum and will scratch the surface of the agar. Streak plates can be made from a broth culture, an agar slant or from an agar plate. It is sometimes convenient to suspend a bit of growth from a solid surface in sterile saline and use this as a source of inoculum. Resuspension of colonies or cultures grown on solid surfaces dilutes the culture and makes streak plating easier.

A loopful of inoculum is transferred from the source and put on the agar surface. When using a large inoculum (a turbid culture or growth from a solid surface), a small spot is spread during the initial transfer. If the inoculum is from a lightly turbid suspension, the first phase of the streaking pattern is begun. The three-phase streaking pattern is recommended for beginners because it is most likely to give satisfactory results with suspensions having a wide range of microbial density.

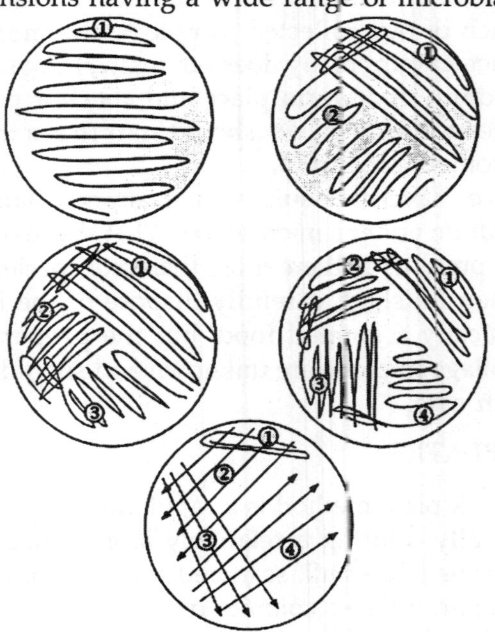

Fig. Streaking Patterns

There are a number of different methods for mechanically diluting microbes on a streak plate. The most common method is spreading microbes across a plate As the concentration of microbes increases so do the number of phases Irrespective of the number of phases loop is flamed between each one. The fifth plate shows an alternative method, where the streaks are not continuous, but are a series of parallel lines. Foobar.

Choosing a streaking pattern is a matter of individual preference and depends upon the number of microorganisms in the sample. Demonstrates the most common patterns, but they are not the only methods. The object of any streaking pattern is the continuous dilution of the inoculum to give many well isolated colonies. For multi-phase streaking it is crucial to flame the loop before starting the next phase. Note the slight overlap into the previous phase to pick up a small inoculum. To streak a plate...

- Flame the loop to sterilize it and let cool.
- Position the plate so that the spot of inoculum is nearest the hand not holding the loop (the opposite hand).
- Lift the plate lid with the opposite hand; just enough to get the loop inside and touch the loop to the inoculum spot. It is often helpful to treat the inoculating loop as if it were a pencil - steadying the loop by resting the heel of the hand against the lab bench.
- Move the loop back and forth across the spot and then gradually continue toward the centre of the plate as we sweep back and forth. Use a very gentle and even pressure.
- When creating each phase, do not worry about keeping each pass across the plate separate from previous ones.
- When about 30% of the plate has been covered by the first streaking phase, remove the loop and flame sterilize it.
- Repeat the above procedure for the second phase, but this time pick up some inoculum by crossing into the

first phase 2-3 times and then not passing into it again.

- Repeat as necessary for the third and fourth phases. After streaking the plate, flame sterilize the loop before setting it down.

Spread Plates and Dilution Plating

An absolute requirement for a microbiologist is to be able to determine the concentration of microorganisms in a given sample. Various particle-counting devices, spectrophotometric methods and microscopic techniques have been used to count cells. However, one drawback to these methods is that they count dead as well as living cells.

The most common method of enumerating viable cells is the plate-count method. Diluting microorganisms and placing them into petri plates (or plates) for incubation is another essential technique for working with microorganisms. This method suffers from some problems. First, only those organisms which can grow on the medium and at the temperature and atmospheric conditions of incubation, will divide and develop into colonies. Second, each colony may not represent the progeny from one cell, as two or more cells (those in clusters, chains or otherwise close to one another) can give rise to one colony.

For these reasons the counts obtained from the plate-count method are given as the number of colony-forming units (CFU's) per ml (or gram) rather than the number of cells per ml (or gram). Despite these drawbacks, the plate-count method is a powerful means by which concentrations of viable organisms may be estimated.

Also, if it is desirable to count a specific subgroup of microorganisms in a sample, selective media or special incubation conditions can often be used to encourage the growth of only this class of organism.As microbial quantitation involves the use of pipettes (or micropipettes) in preparing dilutions and inoculating plates, the beginning student in microbiology must become familiar with their use.

Due to the possibility of ingesting pathogens and toxic

liquids, mouth-pipetting is forbidden in the laboratory! Pipettes are filled and subsequently emptied by the use of propipettes or other pipette bulb. Pay close attention to the demonstration of their use. Important Safety Consideration: When fitting the pipette and pipette bulb together, use very gentle pressure!! Do not jam these items together! (Force is usually not the answer, a good general rule to live by in this lab and in life for that matter.) The glass pipette will probably break and possibly cause severe injury. Handle the pipette only at the top inch or so. For volumes of 5 ml of less, micropipettes are often the tool of choice.

Instruments are available that are capable of dispensing 5 ml all the way to less than 1 μl (one one-millionth of a liter). Micropipettes have made it possible to miniaturize many experiments and greatly decrease the cost of running them. They are also easy to use and can dispense volumes quickly, increasing the number of experiments that can be performed in a set amount of time. Micropipettes are the tool of choice for small volumes.

MICROBIAL CULTIVATION

When microorganisms are cultivated in the laboratory, a growth environment called a medium is used. The medium may be purely chemical (a chemically defined medium), or it may contain organic materials, or it may consist of living organisms such as fertilized eggs.Microorganisms growing in or on such a medium form a culture. A culture is considered a pure culture if only one type of organism is present and a mixed culture if populations of different organisms are present. When first used, the culture medium should be sterile, meaning that no form of life is present before inoculation with the microorganism.

General microbial media. For the cultivation of bacteria, a commonly used medium is nutrient broth, a liquid containing proteins, salts and growth enhancers that will support many bacteria. To solidify the medium, an agent such as agar is added. Agar is a polysaccharide that adds no nutrients to a medium, but merely solidifies it. The medium that results is nutrient agar. Many media for microorganisms are complex, reflecting the growth requirements of the microorganisms.

For instance, most fungi require extra carbohydrate and an acidic environment for optimal growth. The medium employed for these organisms is potato dextrose agar, also known as Sabouraud dextrose agar. For protozoa, liquid media are generally required and for rickettsiae and viruses, living tissue cells must be provided for best cultivation.

For anaerobic microorganisms, the atmosphere must be oxygen free. To eliminate the oxygen, the culture media can be placed within containers where carbon dioxide and hydrogen gas are generated and oxygen is removed from the atmosphere. Commercially available products achieve these conditions. Anaerobic chambers can also be used within closed compartments and technicians can manipulate culture media within these chambers. To encourage carbon dioxide formation, a candle can be burned to use up oxygen and replace it with carbon dioxide.

Special microbial media. Certain microorganisms are cultivated in selective media. These media retard the growth of unwanted organisms while encouraging the growth of the organisms desired. For example, mannitol salt agar is selective for staphylococci because most other bacteria cannot grow in its high-salt environment. Another selective medium is brilliant green agar, a medium that inhibits Gram-positive bacteria while permitting Gram-negative organisms such as *Salmonella* species to grow.

Still other culture media are differential media. These media provide environments in which different bacteria can be distinguished from one another. For instance, violet red bile agar is used to distinguish coliform bacteria such as *Escherichia coli* from noncoliform organisms. The coliform bacteria appear as bright pink colonies in this media, while noncoliforms appear a light pink or clear. Certain media are both selective and differential.

For instance, MacConkey agar differentiates lactose-fermenting bacteria from nonlactose-fermenting bacteria while inhibiting the growth of Gram-positive bacteria. Since lactose-fermenting bacteria are often involved in water pollution, they can be distinguished by adding samples of water to MacConkey agar and waiting for growth to appear.

In some cases, it is necessary to formulate an enriched medium. Such a medium provides specific nutrients that encourage selected species of microorganisms to flourish in a mixed sample. When attempting to isolate *Salmonella* species from fecal samples, for instance, it is helpful to place a sample of the material in an enriched medium to encourage *Salmonella* species to multiply before the isolation techniques begin.

In order to work with microorganisms in the laboratory, it is desirable to obtain them in pure cultures. Pure cultures of bacteria can be obtained by spreading bacteria out and permitting the individual cells to form masses of growth called colonies.

One can then pick a sample from the colony and be assured that it contains only one kind of bacteria. Cultivating these bacteria on a separate medium will yield a pure culture. To preserve microbial cultures, they may be placed in the refrigerator to slow down the metabolism taking place. Two other methods are deep-freezing and freeze-drying.

For deep-freezing, the microorganisms are placed in a liquid and frozen quickly at temperatures below –50°C. Freeze-drying (lyophilization) is performed in an apparatus that uses a vacuum to draw water off after the microbial suspension has been frozen. The culture resembles a powder and the microorganisms can be preserved for long periods in this condition.

STERILIZATION

Wet Heat

Wet heat is the most dependable procedure for the destruction of all forms of microbial life. Steam sterilization generally denotes heating in an autoclave employing saturated steam under a pressure of approximately 15 psi to achieve a chamber temperature of at least 121°C (250°F). The critical factors in insuring the reliability of this sterilization method is:

- Proper temperature and time; and
- The complete replacement of the air with steam (i.e. no entrapment of air).

Some autoclaves utilize a steam activated exhaust valve that remains open during the replacement of air by live steam until the steam triggers the valve to close. Others utilize a pre-cycle vacuum to remove air prior to steam introduction.

Physical controls such as pressure gauges and thermometers are widely used but are considered secondary methods of insuring sterilization. The use of appropriate biological indicators at locations throughout the autoclave is considered the best indicator of sterilization. The biological indicator most widely used for wet heat sterilization is Bacillus stearothermophilus spores.

Dry Heat

Dry heat is less efficient than wet heat sterilization and requires longer times and/or higher temperatures. The specific times and temperatures must be determined for each type of material being sterilized. Generous safety factors are usually added to allow for the variables that can influence the efficiency of this method of sterilization. The moisture of the sterilization environment as well as the moisture history of organisms prior to heat exposure appear to affect the efficiency of dry heat sterilization. Higher temperatures and shorter times may be used for heat resistant materials. The heat transfer properties and the spatial relation or arrangement of articles in the load are critical in insuring effective sterilization.

The advantage of wet heat is a better heat transfer to and into the cell resulting in overall shorter exposure time and lower temperature. Steam sterilization uses pressurized steam at 121-132° C (250-270° F) for 30 or 40 minutes. This type of heat kills all microbial cells including spores, which are normally heat resistant. In order to accomplish the same effect with dry heat in an oven, the temperature needs to be increased to 160-170° C (320-338° F) for periods of 2 to 4 hours.

General Procedures

All materials, equipment, apparatus contaminated with or containing potentially hazardous organisms should be steam sterilized before being washed and stored, or discarded. Autoclaving is the preferred method. Each individual working with biohazardous materials is responsible for sterilization of materials before disposal.

Biohazardous materials should not be placed in autoclaves overnight in anticipation of autoclaving the next day. To minimize hazard to firemen or disaster crews, all biohazardous materials must be placed in an appropriately marked refrigerator or incubator, sterilized, or otherwise confined at the close of each work day.

All autoclaves must be certified for operating efficiency by the periodic use of biological indicator controls and records maintained for three years. Special precautions should be taken to prevent accidental removal of material from an autoclave before it has been sterilized or the simultaneous opening of both doors on a double door autoclave. Dry hypochlorites, or any other strong oxidizing material, must not be autoclaved with organic materials such as a paper, cloth, or oil:

Oxidizer + Organic material + Heat = Possible Explosion.

All floors, laboratory benches and other surfaces in buildings where biohazardous materials are handled should be disinfected as often as deemed necessary by the supervisor. The surroundings should be disinfected after completion of operations involving plating, pipetting, centrifuging and similar procedures with biohazardous materials. It is the responsibility of the supervisor to determine that the disinfectant and the time and method of exposure is effective against the biological agent(s) used in the facility.

Floor drains should be flooded with water or disinfectant at least once each week in order to fill traps and thus prevent the back flow of sewer gases. Floor cleaning procedures which minimize the generation of aerosols should be used. Wet mopping or wet vacuum pick up is recommended.

Water used to mop floors should contain a disinfectant

or disinfectant-detergent. (Dry mopping or dusting should be avoided). Where wet procedures are not practical, dry vacuum cleaning with a HEPA filter on the exhaust, sweeping compound used with push brooms, or dry dust mop heads treated to suppress aerosolization may be used. Stock solutions of suitable disinfectants will be maintained in each laboratory for disinfection purposes. NOTE: working dilutions of certain disinfectants have short shelf lives and must be prepared on a regular basis.

Sterilization Procedures

General criteria for sterilization of typical materials are presented below. Supervisors are encouraged to review the type of materials being handled and to establish standard conditions for sterilization. Treatment conditions to achieve sterility will vary in relation to the volume of material treated, volume of the autoclave, the contamination level, the moisture content and other factors.

Steam Autoclave

- *Laundry:* 250°F (121°C) for a minimum of 30 min.
- *Trash:* 250°F (121°C) for at least 45 minutes per bag. Size of the autoclave and of the bags greatly effect sterilization time. Large bags in a small autoclave may require 90 minutes or more.
- *Glassware:* 250°F (121°C) for a minimum of 25 min.
- *Liquids:* 250°F (121°C) for 25 minutes for each gallon.
- *Animals & bedding:* Steam autoclaving not recommended (sterilization time required would be at least 8 hours). Incineration in an approved facility is the recommended treatment of these wastes.

MICROSCOPES

Since microorganisms are invisible to the unaided eye, the essential tool in microbiology is the microscope. One of the first to use a microscope to observe microorganisms was Robert Hooke, the English biologist who observed algae and fungi in the 1660s. In the 1670s, Anton van Leeuwenhoek, a Dutch merchant, constructed a number of simple microscopes and observed details of numerous forms of protozoa, fungi and bacteria.

During the 1700s microscopes were used to further elaborate on the microbial world and by the late 1800s, the sophisticated light microscopes had been developed. The electron microscope was developed in the 1940s, thus making the viruses and the smallest bacteria (for example, rickettsiae and chlamydiae) visible. Microscopes permit extremely small objects to be seen, objects measured in the metric system in micrometers and nanometers.

A micrometer (ìm) is equivalent to a millionth of a meter, while a nanometer (nm) is a billionth of a meter. Bacteria, fungi, protozoa and unicellular algae are normally measured in micrometers, while viruses are commonly measured in nanometers. A typical bacterium such as *Escherichia coli* measures about two micrometers in length and about one micrometer in width.

Optical Microscopy

Optical or light microscopy involves passing visible light transmitted through or reflected from the sample through a single or multiple lenses to allow a magnified view of the sample. The resulting image can be detected directly by the eye, imaged on a photographic plate or captured digitally. The single lens with its attachments, or the system of lenses and imaging equipment, along with the appropriate lighting equipment, sample stage and support, makes up the basic light microscope.

Limitations of Optical Microscopy

Limitations of standard optical microscopy (bright field microscopy) lie in three areas;

- The technique can only image dark or strongly refracting objects effectively.
- Diffraction limits resolution to approximately 0.2 micrometre.
- Out of focus light from points outside the focal plane reduces image clarity.

Live cells in particular generally lack sufficient contrast to be studied successfully, internal structures of the cell are colourless and transparent. The most common way to increase contrast is to stain the different structures with selective dyes, but this involves killing and fixing the sample. Staining may also introduce artifacts, apparent structural details that are caused by the processing of the specimen and are thus not a legitimate feature of the specimen.

These limitations have, to some extent, all been overcome by specific microscopy techniques which can non-invasively increase the contrast of the image. In general, these techniques make use of differences in the refractive index of cell structures. It is comparable to looking through a glass window: we (bright field microscopy) don't see the glass but merely the dirt on the glass.

There is however a difference as glass is a more dense material and this creates a difference in phase of the light passing through. The human eye is not sensitive to this difference in phase but clever optical solutions have been thought out to change this difference in phase into a difference in amplitude (light intensity).

OPTICAL MICROSCOPY TECHNIQUES

Bright Field Optical Microscopy

Bright field microscopy is the simplest of all the light microscopy techniques. Sample illumination is via transmitted white light, i.e. illuminated from below and observed from above. Limitations include low contrast of most biological samples and low apparent resolution due to the blur of out of focus material. The simplicity of the technique and the minimal sample preparation required are significant advantages.

Oblique Illumination and what it Means

The use of oblique (from the side) illumination gives the image a 3-dimensional appearance and can highlight otherwise invisible features. A more recent technique based on this method is Hoffmann's modulation contrast, a system found on inverted microscopes for use in cell culture. Oblique illumination suffers from the same limitations as bright field microscopy (low contrast of many biological samples; low apparent resolution due to out of focus objects), but may highlight otherwise invisible structures.

Dark field Optical Microscopy

Dark field microscopy is a technique for improving the contrast of unstained, transparent specimens. Darkfield illumination uses a carefully aligned light source to minimise the quantity of directly-transmitted (un-scattered) light entering the image plane, collecting only the light scattered by the sample. Darkfield can dramatically improve image contrast—especially of transparent objects—while requiring little equipment setup or sample preparation. However, the technique does suffer from low light intensity in final image of many biological samples and continues to be affected by low apparent resolution.

Rheinberg illumination is a special variant of dark field illumination in which transparent, colored filters are inserted just before the condenser so that light rays at high aperture are differently colored than those at low aperture (i.e. the background to the specimen may be blue while the object appears self-luminous yellow). Other colour combinations are possible but their effectiveness is quite variable.

Phase Contrast Optical Microscopy

More sophisticated techniques will show differences in optical density in proportion. Phase contrast is a widely used technique that shows differences in refractive index as difference in contrast. It was developed by the Dutch physicist Frits Zernike in the 1930s The nucleus in a cell for example will show up darkly against the surrounding cytoplasm. Contrast is excellent; however it is not for use with thick objects. Frequently, a halo is formed even around small objects, which obscures detail.

The system consists of a circular annulus in the condenser which produces a cone of light. This cone is superimposed on a similar sized ring within the phase-objective. Every objective has a different size ring, so for every objective another condenser setting has to be chosen. The ring in the objective has special optical properties: it first of all reduces the direct light in intensity, but more importantly, it creates an artificial phase difference of about a quarter wavelength. As the physical properties of this direct light have changed, interference with the diffracted light occurs, resulting in the phase contrast image.

Differential Interference Contrast Microscopy

Superior and much more expensive is the use of interference contrast. Differences in optical density will show up as differences in relief. A nucleus within a cell will actually show up as a globule in the most often used differential interference contrast system according to Georges Nomarski.

However, it has to be kept in mind that this is an optical effect and the relief does not necessarily resemble the true shape! Contrast is very good and the condenser aperture can be used fully open, thereby reducing the depth of field and maximizing resolution.

The system consists of a special prism in the condenser that splits light in an ordinary and an extraordinary beam. The spatial difference between the two beams is minimal (less than the maximum resolution of the objective). After passage through the specimen, the beams are reunited by a similar prism in the objective.

In a homogeneous specimen, there is no difference between the two beams and no contrast is being generated. However, near a refractive boundary (say a nucleus within the cytoplasm), the difference between the ordinary and the extraordinary beam will generate a relief in the image. Differential interference contrast requires a polarized light source to function; two polarizing filters have to be fitted in the light path, one below the condenser (the polarizer) and the other above the objective (the analyzer).

Note: In cases where the optical design of a microscope produces an appreciable lateral separation of the two beams we have the case of classical interference microscopy, which does not result in relief images, but can nevertheless be used for the quantitative determination of mass-thicknesses of microscopic objects.

FLUORESCENCE MICROSCOPY

When certain compounds are illuminated with high energy light, they then emit light of a different, lower frequency. This effect is known as fluorescence. Often specimens show their own characteristic autofluorescence image, based on their chemical makeup.

This method is of critical importance in the modern life sciences, as it can be extremely sensitive, allowing the detection of single molecules. Many different fluorescent dyes can be used to stain different structures or chemical compounds. One particularly powerful method is the combination of antibodies coupled to a fluorochrome as in immunostaining. Examples of commonly used fluorochromes are fluorescein or rhodamine. The antibodies can be made tailored specifically for a chemical compound.

For example, one strategy often in use is the artificial production of proteins, based on the genetic code (DNA). These proteins can then be used to immunize rabbits, which then form antibodies which bind to the protein. The antibodies are then coupled chemically to a fluorochrome and then used to trace the proteins in the cells under study.

Highly-efficient fluorescent proteins such as the green fluorescent protein (GFP) have been developed using the molecular biology technique of gene fusion, a process which links the expression of the fluorescent compound to that of the target protein. Piston DW, Patterson GH, Lippincott-Schwartz J, Claxton NS, Davidson MW Nikon MicroscopyU: Introduction to Fluorescent Proteins. Nikon MicroscopyU. Retrieved on 2007-08-22.

This combined fluorescent protein is generally non-toxic to the organism and rarely interferes with the function of the protein under study. Genetically modified cells or organisms directly express the fluorescently-tagged proteins, which enables the study of the function of the original protein in vivo.

Since fluorescence emission differs in wavelength (colour) from the excitation light, a fluorescent image ideally only shows the structure of interest that was labelled with the fluorescent dye. This high specificity led to the widespread use of fluorescence light microscopy in biomedical research. Different fluorescent dyes can be used to stain different biological structures, which can then be detected simultaneously, while still being specific due to the individual colour of the dye.

To block the excitation light from reaching the observer or the detector, filter sets of high quality are needed. These typically consist of an excitation filter selecting the range of excitation wavelengths, a dichroic mirror and an emission filter blocking the excitation light.

Most fluorescence microscopes are operated in the Epi-illumination mode to further decrease the amount of excitation light entering the detector.

Confocal Laser Scanning Microscopy

Generates the image by a completely different way than the normal visual bright field microscope. It gives slightly higher resolution, but most importantly it provides optical sectioning without disturbing out-of-focus light degrading the image. Therefore it provides sharper images of 3D objects. This is often used in conjunction with fluorescence microscopy.

Deconvolution Microscopy

Fluorescence microscopy is extremely powerful due to its ability to show specifically labelled structures within a complex environment but also because of its inherent ability to provide three dimensional information of biological structures. Unfortunately this information is blurred by the fact, that upon illumination all fluorescently labeled structures emit light no matter if they are in focus or not.

This means, that an image of a certain structure is always blurred by the contribution of light from structures which are out of focus. This phenomenon becomes apparent as a loss of contrast especially when using objectives with a high resolving power, typically oil immersion objectives with a high numerical aperture.

Fortunately though, this phenomenon is not caused by random processes such as light scattering but can be relatively well defined by the optical properties of the image formation in the microscope imaging system. If one considers a small fluorescent light source (essentially a bright spot), light coming from this spot spreads out the further out of focus one is. Under ideal conditions this produces a sort of "hourglass" shape of this point source in the third (axial) dimension. This shape is called the point spread function (PSF) of the microscope imaging system

Since any fluorescence image is made up of a large number of such small fluorescent light sources the image is said to be "convolved by the point spread function".

Knowing this point spread function means, that it is possible to reverse this process to a certain extent by computer based methods commonly known as deconvolution microscopy. There are various algorithms available for 2D or 3D Deconvolution. They can be roughly classified in non restorative and restorative methods.

While the non restorative methods can improve contrast by removing out of focus light from focal planes, only the restorative methods can actually reassign light to it proper place of origin. This can be an advantage over other types of 3D microscopy such as confocal microscopy, because light is not thrown away but reused. For 3D deconvolution one typically provides a series of images derived from different focal planes (called a Z-stack) plus the knowledge of the PSF which can be either derived experimentally or theoretically from knowing all contributing parameters of the microscope.

Sub-diffraction Optical Microscopy Techniques

It is well known that there is a spatial limit to which light can focus: approximately half of the wavelength of the light we are using. But this is not a true barrier, because this diffraction limit is only true in the far-field and localization precision can be increased with many photons and careful analysis (although two objects still cannot be resolved); and like the sound barrier, the diffraction barrier is breakable.

This section explores some approaches to imaging objects smaller than ~250 nm. Most of the following information was gathered (with permission) from a chemistry blog's review of sub-diffraction microscopy techniques.

TYPES OF MICROSCOPES

Various types of microscopes are available for use in the microbiology laboratory. The microscopes have varied applications and modifications that contribute to their usefulness.The light microscope. The common light microscope used in the laboratory is called a compound microscope because it contains two types of lenses that function to magnify an object.

The lens closest to the eye is called the ocular, while the lens closest to the object is called the objective. Most microscopes have on their base an apparatus called a condenser, which condenses light rays to a strong beam. A diaphragm located on the condenser controls the amount of light coming through it. Both coarse and fine adjustments are found on the light microscope.

To magnify an object, light is projected through an opening in the stage, where it hits the object and then enters the objective. An image is created and this image becomes an object for the ocular lens, which remagnifies the image. Thus, the total magnification possible with the microscope is the magnification achieved by the objective multiplied by the magnification achieved by the ocular lens.

A compound light microscope often contains four objective lenses: the scanning lens (4X), the low-power lens (10X), the high-power lens (40 X) and the oil-immersion lens (100 X). With an ocular lens that magnifies 10 times, the total magnifications possible will be 40 X with the scanning lens, 100 X with the low-power lens, 400 X with the high-power lens and 1000 X with the oil-immersion lens. Most microscopes are parfocal. This term means that the microscope remains in focus when one switches from one objective to the next objective.

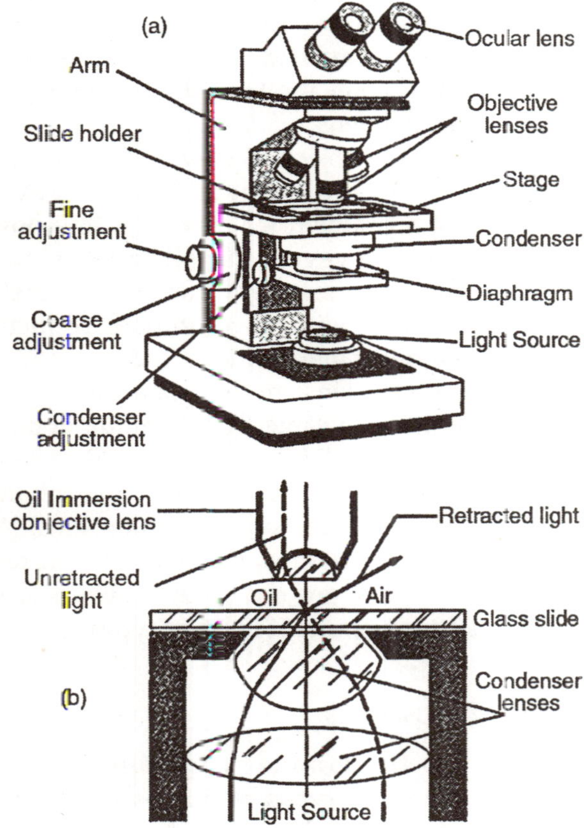

Fig. Light Microscopy. (a) The Important Parts of a Common Light Microscope. (b) How Immersion oil Gathers more Light for use in the Microscope

The ability to see clearly two items as separate objects under the microscope is called the resolution of the microscope. The resolution is determined in part by the wavelength of the light used for observing. Visible light has a wavelength of about 550 nm, while ultraviolet light has a wavelength of about 400 nm or less. The resolution of a microscope increases as the wavelength decreases, so ultraviolet light allows one to detect objects not seen with visible light. The resolving power of a lens refers to the size of the smallest object that can be seen with that lens. The resolving power is based on the wavelength of the light used and the numerical aperture of the lens. The numerical aperture (NA) refers to the widest cone of light that can enter the lens; the NA is engraved on the side of the objective lens.

If the user is to see objects clearly, sufficient light must enter the objective lens. With modern microscopes, entry to the objective is not a problem for scanning, low-power and high-power lenses. However, the oil-immersion lens is exceedingly narrow and most light misses it. Therefore, the object is seen poorly and without resolution. To increase the resolution with the oil-immersion lens, a drop of immersion oil is placed between the lens and the glass slide.

Immersion oil has the same light-bending ability (index of refraction) as the glass slide, so it keeps light in a straight line as it passes through the glass slide to the oil and on to the glass of the objective, the oil-immersion lens. With the increased amount of light entering the objective, the resolution of the object increases and one can observe objects as small as bacteria. Resolution is important in other types of microscopy as well.

Other Light Microscopes

In addition to the familiar compound microscope, microbiologists use other types of microscopes for specific purposes. These microscopes permit viewing of objects not otherwise seen with the light microscope. An alternative microscope is the dark-field microscope, which is used to observe live spirochetes, such as those that cause syphilis. This microscope contains a special condenser that scatters light and causes it to reflect off the specimen at an angle. A light object is seen on a dark background.

A second alternative microscope is the phase-contrast microscope. This microscope also contains special condensers that throw light "out of phase" and cause it to pass through the object at different speeds. Live, unstained organisms are seen clearly with this microscope and internal cell parts such as mitochondria, lysosomes and the Golgi body can be seen with this instrument. The fluorescent microscope uses ultraviolet light as its light source.

When ultraviolet light hits an object, it excites the electrons of the object and they give off light in various shades of colour. Since ultraviolet light is used, the resolution of the object increases. A laboratory technique called the fluorescent-antibody technique employs fluorescent dyes and antibodies to help identify unknown bacteria.

Electron Microscopy

The energy source used in the electron microscope is a beam of electrons. Since the beam has an exceptionally short wavelength, it strikes most objects in its path and increases the resolution of the microscope significantly. Viruses and some large molecules can be seen with this instrument. The electrons travel in a vacuum to avoid contact with deflecting air molecules and magnets focus the beam on the object to be viewed. An image is created on a monitor and viewed by the technologist. The more traditional form of electron microscope is the transmission electron microscope (TEM).

To use this instrument, one places ultrathin slices of microorganisms or viruses on a wire grid and then stains them with gold or palladium before viewing. The densely coated parts of the specimen deflect the electron beam and both dark and light areas show up on the image. The scanning electron microscope (SEM) is the more contemporary form electron microscope. Although this microscope gives lower magnifications than the TEM, the SEM permits three-dimensional views of microorganisms and other objects. Whole objects are used and gold or palladium staining is employed.

Staining Techniques

Because microbial cytoplasm is usually transparent, it is necessary to stain microorganisms before they can be viewed with the light microscope. In some cases, staining is unnecessary, for example when microorganisms are very large or when motility is to be studied and a drop of the microorganisms can be placed directly on the slide and observed. A preparation such as this is called a wet mount.

A wet mount can also be prepared by placing a drop of culture on a cover-slip (a glass cover for a slide) and then inverting it over a hollowed-out slide. This procedure is called the hanging drop.

In preparation for staining, a small sample of microor ganisms is placed on a slide and permitted to air dry. The smear is heat fixed by quickly passing it over a flame. Heat fixing kills the organisms, makes them adhere to the slide and permits them to accept the stain.

Simple Stain Techniques

Staining can be performed with basic dyes such as crystal violet or methylene blue, positively charged dyes that are attracted to the negatively charged materials of the microbial cytoplasm. Such a procedure is the simple stain procedure. An alternative is to use a dye such as nigrosin or Congo red, acidic, negatively charged dyes. They are repelled by the negatively charged cytoplasm and gather around the cells, leaving the cells clear and unstained. This technique is called the negative stain technique.

Differential Stain Techniques

The differential stain technique distinguishes two kinds of organisms. An example is the Gram stain technique. This differential technique separates bacteria into two groups, Gram-positive bacteria and Gram-negative bacteria. Crystal violet is first applied, followed by the mordant iodine, which fixes the stain.

Then the slide is washed with alcohol and the Gram-positive bacteria retain the crystal-violet iodine stain; however, the Gram-negative bacteria lose the stain. The Gram-negative bacteria subsequently stain with the safranin dye, the counterstain, used next. These bacteria appear red under the oil-immersion lens, while Gram-positive bacteria appear blue or purple, reflecting the crystal violet retained during the washing step.

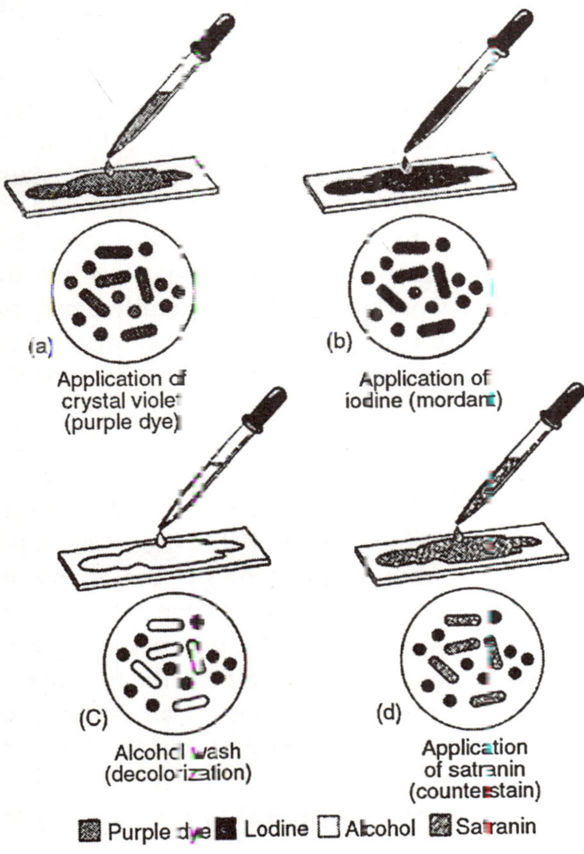

Fig. The Gram Stain Procedure used for Differentiating Bacteria into two Groups

Another differential stain technique is the acid-fast technique. This technique differentiates species of *Mycobacterium* from other bacteria. Heat or a lipid solvent is used to carry the first stain, carbolfuchsin, into the cells. Then the cells are washed with a dilute acid-alcohol solution. *Mycobacterium* species resist the effect of the acid-alcohol and retain the carbolfuchsin stain (bright red). Other bacteria lose the stain and take on the subsequent methylene blue stain (blue). Thus, the acid-fast bacteria appear bright red, while the nonacid-fast bacteria appear blue when observed under oil-immersion microscopy.

Other stain techniques seek to identify various bacterial structures of importance. For instance, a special stain technique highlights the flagella of bacteria by coating the flagella with dyes or metals to increase their width. Flagella so stained can then be observed.

A special stain technique is used to examine bacterial spores. Malachite green is used with heat to force the stain into the cells and give them colour. A counterstain, safranin, is then used to give colour to the nonsporeforming bacteria. At the end of the procedure, spores stain green and other cells stain red.

LIGHT MICROSCOPY

The light microscope, so called because it employs visible light to detect small objects, is probably the most well-known and well-used research tool in biology. Yet, many students and teachers are unaware of the full range of features that are available in light microscopes. Since the cost of an instrument increases with its quality and versatility, the best instruments are, unfortunately, unavailable to most academic programs. However, even the most inexpensive "student" microscopes can provide spectacular views of nature and can enable students to perform some reasonably sophisticated experiments.

A beginner tends to think that the challenge of viewing small objects lies in getting enough magnification. In fact, when it comes to looking at living things the biggest challenges are, in order,

- Obtaining sufficient contrast
- Finding the focal plane
- Obtaining good resolution
- Recognizing the subject when one sees it

The smallest objects that are considered to be living are

the bacteria. The smallest bacteria can be observed and cell shape recognized at a mere 100x magnification. They are invisible in bright field microscopes, though. These pages will describe types of optics that are used to obtain contrast, suggestions for finding specimens and focusing on them and advice on using measurement devices with a light microscope.

Types of Light Microscopes

The bright field microscope is best known to students and is most likely to be found in a classroom. Better equipped classrooms and labs may have dark field and/or phase contrast optics. Differential interference contrast, Nomarski, Hoffman modulation contrast and variations produce considerable depth of resolution and a three dimensional effect. Fluorescence and confocal microscopes are specialized instruments, used for research, clinical and industrial applications.

Other than the compound microscope, a simpler instrument for low magnification use may also be found in the laboratory. The stereo microscope, or dissecting microscope usually has a binocular eyepiece tube, a long working distance and a range of magnifications typically from 5x to 35 or 40x. Some instruments supply lenses for higher magnifications, but there is no improvement in resolution. Such "false magnification" is rarely worth the expense.

Bright Field Microscopy

With a conventional bright field microscope, light from an incandescent source is aimed toward a lens beneath the stage called the condenser, through the specimen, through an objective lens and to the eye through a second magnifying lens, the ocular or eyepiece. We see objects in the light path because natural pigmentation or stains absorb light differentially, or because they are thick enough to absorb a significant amount of light despite being colorless.

A *Paramecium* should show up fairly well in a bright field microscope, although it will not be easy to see cilia or most organelles. Living bacteria won't show up at all unless the viewer hits the focal plane by luck and distorts the image by using maximum contrast. A good quality microscope has a built-in illuminator, adjustable condenser with aperture diaphragm (contrast) control, mechanical stage and binocular eyepiece tube.

The condenser is used to focus light on the specimen through an opening in the stage. After passing through the specimen, the light is displayed to the eye with an apparent field that is much larger than the area illuminated. The magnification of the image is simply the objective lens magnification (usually stamped on the lens body) times the ocular magnification.

Students are usually aware of the use of the coarse and fine focus knobs, used to sharpen the image of the specimen. They are frequently unaware of adjustments to the condenser that can affect resolution and contrast. Some condensers are fixed in position, others are focusable, so that the quality of light can be adjusted. Usually the best position for a focusable condenser is as close to the stage as possible.

The bright field condenser usually contains an aperture diaphragm, a device that controls the diameter of the light beam coming up through the condenser, so that when the diaphragm is stopped down (nearly closed) the light comes straight up through the centre of the condenser lens and contrast is high. When the diaphragm is wide open the image is brighter and contrast is low.

A disadvantage of having to rely solely on an aperture diaphragm for contrast is that beyond an optimum point the more contrast we produce the more we distort the image. With a small, unstained, unpigmented specimen, we are usually past optimum contrast when we begin to see the image.

Mount the Specimen on the Stage

The cover slip must be up if there is one. High magnification objective lenses can't focus through a thick glass slide; they must be brought close to the specimen, which is why coverslips are so thin. The stage may be equipped with simple clips (less expensive microscopes), or with some type of slide holder. The slide may require manual positioning, or there may be a mechanical stage (preferred) that allows precise positioning without touching the slide.

Optimize the Lighting

A light source should have a wide dynamic range, to provide high intensity illumination at high magnifications and lower intensities so that the user can view comfortably at low magnifications.

Better microscopes have a built-in illuminator and the best microscopes have controls over light intensity and shape of the light beam. If our microscope requires an external light source, make sure that the light is aimed toward the middle of the condenser. Adjust illumination so that the field is bright without hurting the eyes.

Adjust the Condenser

To adjust and align the microscope, start by reading the manual. If no manual is available, try using these guidelines. If the condenser is focusable, position it with the lens as close to the opening in the stage as we can get it. If the condenser has selectable options, set it to bright field. Start with the aperture diaphragm stopped down (high contrast). We should see the light that comes up through the specimen change brightness as we move the aperture diaphragm lever.

FOCUS, LOCATE AND CENTRE THE SPECIMEN

Start with the lowest magnification objective lens, to home in on the specimen and/or the part of the specimen we wish to examine. It is rather easy to find and focus on sections of tissues, especially if they are fixed and stained, as with most prepared slides. However it can be very difficult to locate living, minute specimens such as bacteria or unpigmented protists. A suspension of yeast cells makes a good practice specimen for finding difficult objects.

- Use dark field mode (if available) to find unstained specimens. If not, start with high contrast (aperture diaphragm closed down).

- Start with the specimen out of focus so that the stage and objective must be brought closer together. The first surface to come into focus as we bring stage and objective together is the top of the cover slip. With smears, a cover slip is frequently not used.
- If we are having trouble, focus on the edge of the cover slip or an air bubble, or something that we can readily recognize. The top edge of the cover slip comes into focus first, then the bottom, which should be in the same plane as our specimen.
- Once we have found the specimen, adjust contrast and intensity of illumination and move the slide around until we have a good area for viewing.

Adjust Eyepiece Separation, Focus

With a single ocular, there is nothing to do with the eyepiece except to keep it clean. With a binocular microscope (preferred) we need to adjust the eyepiece separation just like we do a pair of binoculars. Binocular vision is much more sensitive to light and detail than monocular vision, so if we have a binocular microscope, take advantage of it.

One or both of the eyepieces may be a telescoping eyepiece, that is, we can focus it. Since very few people have eyes that are perfectly matched, most of us need to focus one eyepiece to match the other image. Look with the appropriate eye into the fixed eyepiece and focus with the microscope focus knob. Next, look into the adjustable eyepiece (with the other eye of course) and adjust the eyepiece, not the microscope.

Select an Objective lens for Viewing

The lowest power lens is usually 3.5 or 4x and is used primarily for initially finding specimens. We sometimes call it the scanning lens for that reason. The most frequently used objective lens is the 10x lens, which gives a final magnification of 100x with a 10x ocular lens.

For very small protists and for details in prepared slides such as cell organelles or mitotic. Typical high magnification lenses are 40x and 97x or 100x. The latter two magnifications are used exclusively with oil in order to improve resolution. Move up in magnification by steps. Each time we go to a higher power objective, re-focus and re-centre the specimen. Higher magnification lenses must be physically closer to the specimen itself, which poses the risk of jamming the objective into the specimen.

Be very cautious when focusing. By the way, good quality sets of lenses are parfocal, that is, when we switch magnifications the specimen remains in focus or close to focused.

Bigger is not always better. All specimens have three dimensions and unless a specimen is extremely thin we will be unable to focus with a high magnification objective. The higher the magnification, the harder it is to "chase" a moving specimen.

The apparent field of an eyepiece is constant regardless of magnification used. So it follows that when we raise magnification the area of illuminated specimen we see is smaller. Since we are looking at a smaller area, less light reaches the eye and the image darkens. With a low power objective we may have to cut down on illumination intensity. With a high power we need all the light we can get, especially with less expensive microscopes.

Bright field microscopy is best suited to viewing stained or naturally pigmented specimens such as stained prepared slides of tissue sections or living photosynthetic organisms. It is useless for living specimens of bacteria and inferior for non-photosynthetic protists or metazoans, or unstained cell suspensions or tissue sections. Here is a not-so-complete list of specimens that might be observed using bright-field microscopy and appropriate magnifications (preferred final magnifications are emphasized).

- *Prepared slides, stained:* Bacteria (1000x), thick tissue sections (100x, 400x), thin sections with condensed chromosomes or specially stained organelles (1000x), large protists or metazoans (100x).
- *Smears, stained:* Blood (400x, 1000x), negative stained bacteria (400x, 1000x).

- *Living preparations (wet mounts, unstained):* Pond water (40x, 100x, 400x), living protists or metazoans (40x, 100x, 400x occasionally), algae and other microscopic plant material (40x, 100x, 400x). Smaller specimens will be difficult to observe without distortion, especially if they have no pigmentation.

Care of the Microscope

- Everything on a good quality microscope is unbelievably expensive, so be careful.
- Hold a microscope firmly by the stand, only. Never grab it by the eyepiece holder, for example.
- Hold the plug (not the cable) when unplugging the illuminator.
- Since bulbs are expensive and have a limited life, turn the illuminator off when we are done.
- Always make sure the stage and lenses are clean before putting away the microscope.
- Never use a paper towel, a kimwipe, our shirt, or any material other than good quality lens tissue or a cotton swab (must be 100% natural cotton) to clean an optical surface. We may use an appropriate lens cleaner or distilled water to help remove dried material. Organic solvents may separate or damage the lens elements or coatings.
- Cover the instrument with a dust jacket when not in use.
- Focus smoothly; don't try to speed through the focusing process or force anything. For example if we encounter increased resistance when focusing then we 've probably reached a limit and we are going in the wrong direction.

Chapter 3

Tools in Nanobiology

ATOMIC FORCE MICROSCOPY

Genomics have contributed greatly to the understanding of the molecular basis of disease and to the development of new therapies and drug design. In the post-genome era, proteomics aim to translate the nucleic acid information archive into an understanding of how the cell actually works and how disease processes operate.

The most relevant information archive to clinical applications and drug development involves the elucidation of the information flow of the cell: the programme of protein pathway networks. There are several reasons to use the atomic force microscope (AFM) in the field of proteomics.

First, the AFM provides a detailed look into the overall structure of proteins. Information on size, shape and subunit composition of the protein can be recorded and, on a functional base, it is possible to follow molecular interactions, *e.g.* bio-assembly processes.

Second, protein surface interactions, the protein surface binding process and how it affects its biological functionality, can be investigated by AFM techniques. The results can be used for the development of new biosensors and immunoassays.

Considering the recent advances in experimental genomics and proteomics, we emphasize the use of AFM as a force spectroscopy tool. When used as a sensor, AFM opens the possibility of enormous change in our ability to analyse and interpret complex biological processes, allowing, for example, the detection and elucidation of protein-DNA and protein-protein interactions.

AFM AS A SENSOR

Understanding the force (derivative of energy with respect to distance) that drives specific molecular interactions is a challenging task in molecular and structural biology, because it is the most direct means of obtaining information about how the interaction energy between two different structures is distributed in space.

Such specific interactions result in multiple weak, non-covalent bonds formed between defined portions of the interacting molecular partners. The development of the Atomic Force Microscope (AFM), as a Scanning Probe Microscopy (SPM) family member, has opened new perspectives for the investigation of surfaces at high lateral and vertical resolution.

During the last decade, AFM was proposed to study inter- and intramolecular interactions forces in biological macromolecules, mainly due its precision and sensitivity to probe surfaces in physiological environments with molecular resolution and with forces down to the pico-newton range. Thus AFM is a useful tool to evaluate and to characterize mechanical properties of biological samples such as: topography, elasticity and adhesive properties.

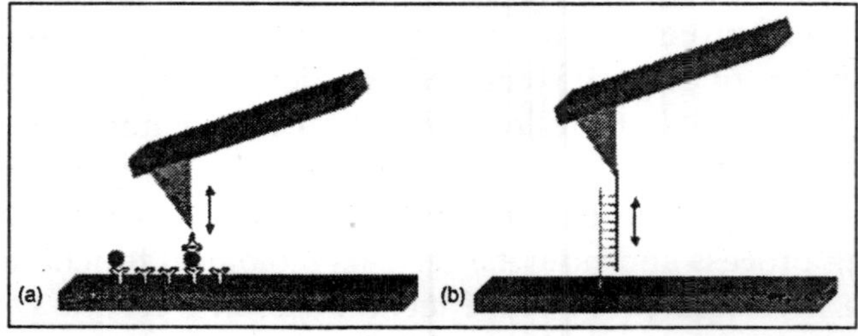

Fig. Schematic set up of force spectroscopy experiments: By immobilizing a molecular partner either over AFM tip and substrate AFM can test a) intermolecular interactions between receptor and ligand; and b) complementary nucleic acid binding between DNA strands.

Furthermore, due to its ability to operate at least as well

in liquid as in air, AFM opens the possibility of probing biological sample interactions in aqueous buffer *i.e.*, under conditions, close to their native environment.

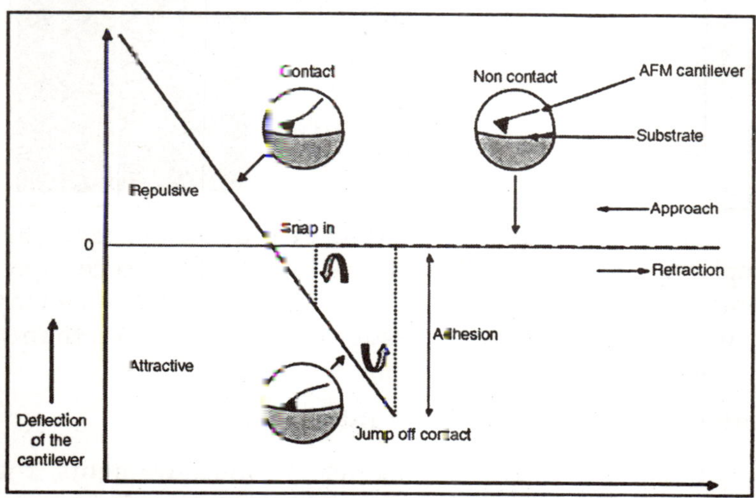

Fig. Force contact curve generated by an AFM: The deflection of the cantilever is shown as a function of the tip-sample surface distance. In the beginning of the approach cycle the AFM tip is not in contact with the surface; then, the AFM tip is being pushed into the surface resulting in a bending of the cantilever. In the retraction cycle the tip is being withdrawn from the surface, the tip adheres to the sample surface and, after the jump-off contact from the surface, the tip is brought to the non-contact position again.

When recording a force curve the AFM tip starts far from the sample surface, where no interactions between either exist. As the tip-sample separation is reduced beyond a certain point, attractive dispersion and electrostatic forces between the two surfaces begin to interact, causing the flexible cantilever to bend towards the sample up to a certain point, where strong repulsive forces bend the cantilever outwards. At each distance, the cantilever bends until its elastic (restoring) force equals the tip-sample interaction force and the system is in equilibrium.

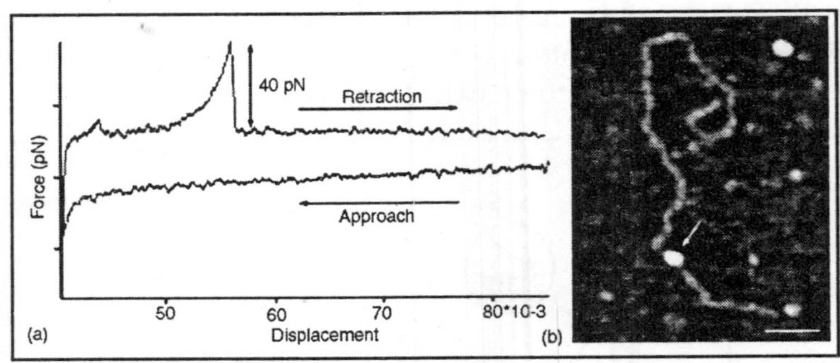

The resultant plot of the cantilever deflection *versus* the separation between the tip and the sample is the force curve. The AFM combines high force sensitivity in the pico-newton range with high lateral resolution, often better than a nanometer, which is in the molecular range.

Inter- and intra-molecular interactions

For quantitative force adhesion measurements, the partners in the molecular recognition reaction have to be immobilized on both the substrate surface and the AFM tip. Furthermore, the specific interactions must be compared to the background of non-specific probe/surface interactions, which may be as high as, or even higher, than the biological interactions.

This is done by using non-functionalized probes or surfaces for comparison, by blocking the interactions between the immobilized molecular partners with free ligands in the medium, or by changing the pH or the salt concentration in the medium.

Concerning intermolecular interactions, biotin/ streptavidin (or avidin) has been used as a model system for ligandreceptor adhesion studies, because of its high affinity and the availability of structural and thermodynamic data. The unbinding forces of antibody-antigen complexes measured by means of AFM force spectroscopy correlate with the spontaneous association (kon) and dissociation (koff), and the dissociation constant (KD= koff/kon), which describes the equilibrium behaviour.

Furthermore, single molecule AFM experiments measuring coupling of stretch and twist movements, overstretching mechanics and base pairing forces provided an intimate look into the biophysical details of DNA. Direct measurements of the unbinding forces between single strands of DNA used covalently bound DNA oligonucleotides to the AFM tip and surface, respectively. The adhesive forces between complementary 20 base pair strands are in the nanonewton range, and the intra chain interaction resulting from the molecule's elasticity showed a long-range cohesive force behaviour.

Sequence-dependent mechanical properties of DNA measured by unzipping experiments showed that the transition from B- to S-DNA conformation occurs at significantly lower forces (65 piconewtons) in poly (dA-dT) compared to poly (dG-dC) oligonucleotides. Using individual double-stranded DNA molecules attached between an AFM tip and a gold surface, the mechanical stability of overstretched DNA was described. In the case of Lambda phage a DNA B-S transition was observed followed by a second conformational transition, during which the DNA double helix melts into two single strands.

A major focus of current genomics research is the detection of single-nucleotide polymorphisms (SNPs) within known gene sequence, as well as in the whole genome. Such mismatched base pairing was detected using a gold-coated AFM cantilever functionalized with thiolated 20- or 25-mer probe DNA oligonucleotides and exposed to target oligonucleotides of varying sequence in static and flow conditions, providing a distinct positive/negative signal for easy interpretation of oligonucleotides hybridization.

The association of SNPs with disorders is expected to be particularly useful in identification of cancer and neural degenerative disorders such as Alzheimer's disease and is also related to the individual variations in drug responses.

Recently, dynamic studies of DNA-protein interactions came into the forefront of biophysical research. Single molecule force spectroscopy showed the specificity of the interaction between ExpG protein and the promoter regions in the galactoglucan biosynthesis (*exp*) gene cluster. Moreover the binding mechanism involved with respect to its thermal off-rate and additional molecular parameters, describing the energy landscape of the separation process, was characterized.

Our group has shown the specific interactions between LexA repressor protein, derived from the SOS repair system of *Escherichia coli*, and two specific DNA binding motifs, *recA* and *yebG*. This experiment showed a pronounced feature in the energy landscape along a force-driven pathway between the rupture force and the off rate (koff) value for each operator.

Furthermore AFM force spectroscopy can be used to acquire detailed information in the three-dimensional structure of chromosomes. These studies are important to understand the mechanisms of histone movement for transcription, replication and repair of DNA. To determine the forces necessary to break the bonds between the DNA and histone octamers in the nucleosomal particles, AFM experiments were carried out by stretching single chromatin fibers bound to the AFM tip.

In the post-genomic area the understanding of protein folding plays an important role and has wide-ranging impact in structural biology. AFM also allows the study of intramolecular interactions like protein-folding mechanisms, especially unfolding of multi-domain protein molecules or individual protein domains.

The unfolding and stretching of each domain creates an individual peak in the force curve, leading to a characteristic saw-tooth pattern. Single molecule force measurements of individual titin immunoglobulin domains showed an unfolding by forces in the range of 150 to 300 pN, and dependent on the pulling speed. After relaxation, refolding of immunoglobulin domains could be observed.

Using recombinant constructs from different titin parts, the distribution of mechanical stability along more than 200 Ig and fibronectin III domains in titin could be demonstrated. It has been shown that the force required for the mechanical unfolding spectrin repeats is between 25 to 35 pN. In other words, the unfolding forces of the helical spectrin domains are five to ten times lower than those found in domains with, fold, like in the immunoglobulin or fibronectin III domains, where the tertiary structure is stabilized by hydrogen bonds between adjacent strands.

This demonstrates that the forces which stabilize the coiled-coil lead to a mechanically much weaker structure than multiple hydrogen-bonded, sheets. The combination of high resolution AFM imaging with force spectroscopy provides an insight into the interaction forces between the individual protomers of the hexagonally packed intermediate (HPI) layer of *Deinococcus radiodurans*.

Individual protomers were sequentially stretched, unfolded and removed from a bacterial surface layer until an entire bacterial pore formed by six protomers was unzipped. Besides the possibility of single molecule force spectroscopy, the AFM can also be used as a manipulator. This was demonstrated with purple membrane patches, from *Halobacterium salinarum*, where individual molecules of the membrane protein bacteriorhodopsin could be localized, unfolded and extracted from the lipid membrane.

Further, the influence of *pH* and local mutations on the stability of individual structural elements of bacteriorhodopsin against mechanical unfolding has been analyzed. The polypeptide loops act as a barrier for unfolding and contribute significantly to the structural stability of bacteriorhodopsin.

Cellular adhesion mediated by biological macromolecules and their respective ligands plays an essential role in a number of diverse biological phenomena including inflammation and cancer metastasis. How the adhesiveness of receptor-ligand interactions is controlled by the affinity of the individual receptors to single ligands is not well understood.

Using single molecule force spectroscopy, the tensile strength and off-rate of single P-selectin molecules binding to single ligands on intact human polymorphonuclear leukocytes and metastatic colon carcinomas were probed *in situ* and compared to the overall adhesiveness of these cells for P-selectin substrates.

Characterizing the biochemical and biophysical properties of functional P-selectin ligands on carcinomas will provide guidelines to engineer novel therapeutic agents that will selectively block ligand function and thus interfere with metastatic spreads.

Force-distance curves have also been utilized to identify cell partners that interact specifically in certain biological reactions. Cell-cell adhesion mediated by specific cell-surface molecules is essential for multi-cellular development. By functionalizing AFM tips with whole cells of a given type and studying their interaction with monolayers of other cell types, it was possible to identify the cell type in the uterine epithelium that interacts specifically with cells in the embryo during implantation.

The cell-cell adhesion mediated by specific cell-surface molecules has been investigated using a glycoprotein layer, contact site A (csA) as a prototype of cell-adhesion proteins. csA is expressed in aggregating cells of *D. discoideum,* which are engaged in development of a multicellular organism.

VISCOELASTIC PROPERTIES OF BIOMOLECULES

AFM has also become a powerful tool to measure the viscoelastic properties of biological structures and macromolecules. Whenever the effective stiffness of the cantilever and the biological sample on the surface are of comparable size and the AFM tip is pushed into the sample, the sample is indented.

At the time when the stress (deformation force) and the strain (the amount of deformation) are linearly related, the deformation of the material is elastic and the material will regain its original form upon relaxation. The depth of indentation can be used to perform local elasticity measurements of Young's elastic modulus (the mechanical resistance of a material while elongating or compressing).

The capability of the AFM to provide information on the elastic properties of biological structures has been used to study different types of differentiated cells and organelles and human chromosomes. The elasticity of human chromosomes was determined in neutral and alkaline pH but also in acetate buffers.

The behaviour of soft biological samples is different to a hard surface, indicative in the force retraction curve. Whereas lift-off occurs quickly on hard surfaces, it may be considerably slowed down in the case of soft biological samples. For example, the lift-off speed has be used to estimate the viscosity of the sample and the elasticity of lysozyme adsorbed on mica.

Force mapping

Individual curves can be assembled into a force-volume providing a three-dimensional, laterally resolved description of the forces within the sample. A force-volume can be used to produce sample/surface maps reflecting different properties of the surface as adhesion, viscosity, elasticity, *etc.* Using this approach a map of antigenic sites on a surface by molecular recognition of an antigen by an antibody tethered to an AFM tip has been acquired. The question as to whether cell division is driven by cortical relaxation outside the equatorial region or by cortical contractibility with the developing furrow alone has been approached by monitoring spatially-resolved changes in the cortical stiffness with time. Here force-mapping was used to track dynamic changes in the stiffness of the cortex of adherent cultured cells along a single scan-line during methaphase to cytokinesis. The Atomic Force Microscope (AFM) as a force spectroscopy tool in genomics and proteomics opened new opportunities for studying the mechanical properties of biomolecules and their interactions in their native environment. Those tools have shown useful applications in genomics and proteomics studies by determining the binding affinity of DNA proteins dependent on the target DNA sequence for further correlative studies on physical affinity and biological relevance of the controlled gene.

Further, force spectroscopy is a powerful analytical tool to investigate structural and functional features of biomolecules. AFM used in programmable DNA sensors and protein biochips opens promising applications in the new field of nanobiotechnology.

Chapter 4

Nanobiology Strategies

UNDERSTANDING THE IMMUNE SYSTEM

Recent advances in science at the nanometer scale have made the concept of using nanotechnology for detection and sampling in vivo a viable proposition. Nanoobjects (e.g. nanotubes, nanofibers, nanoparticles, quantum dots) can provide a wealth of information that cannot be obtained with the same accuracy and sensitivity by any other existing methodology.

Thus, the application of nanotechnology to immunological questions could greatly expand the current understanding of the immune response. Much of the work to date in the field has been focused on aerospace and aviation, not biology or medicine. However, the areas of nanobiology and nanosystems biology are beginning to attract researchers who understand the enormous potential value of applying this technology to immunological problems, targeted drug delivery and increased accuracy and sensitivity in analyzing laboratory samples.

DIAGNOSTICS

Some of the difficulties in detecting the immune response to pathogen invasion are the amount of biological material needed to test for a response, the lack of sensitivity of the test, and the time it takes to process the material. Chad Mirkin, at Northwestern University, has developed a highly sensitive method for detecting protein analytes using nanoparticle probes that are specific to a target of interest. A complex is created from a nanoparticle probe and a magnetic microparticle probe coated with antibodies specific to the target of interest. Prostate specific antigen was the target used in the original study. Dehybridization of the oligonucleotides on the nanoparticle probe surface is performed to determine the presence of the target protein by identifying the oligonucleotides sequence released from the nanoparticle probe, in essence a biobarcode.

Because the nanoparticle probe carries with it a large number of oligonucleotides per protein binding event, there is substantial amplification of the DNA, allowing detection of protein targets in the 500 zeptomolar (10^{-21}M) to picomolar (10^{-12}M) concentration range. Clinically accepted conventional assays for detecting the same targets have sensitivity limits of approximately 3 picomolar. This assay has been developed for prostate cancer, HIV, cardiac markers and Alzheimer's disease, and could be adopted for detection of pathogens used in a bioterrorist attack.

Rashid Bashir, from Purdue University, described his work in creating devices in a lab-on-a-chip format that can be used for rapid detection and characterization of cells and microorganisms for real-time analysis of nucleic acids and proteins. One such system takes advantage of the dielectrophoretic effect to sort and concentrate microorganisms within a micro-fluidic biochip allowing specific capture of particles inside the chip.

Another device is used to detect the mass of bacteria and viruses as they bind to nano-mechanical silicon cantilevers. Single molecule DNA analysis using silicon-based nano-pore channel sensors can be integrated in these lab-on-a-chip devices for direct electrical characterization of these molecules expressed by a specific cell.

THERAPEUTICS

Researchers are developing ways to use nanotechnology to deliver therapies targeted to a pathogen or virus without harm to surrounding tissue and in creating formulations that can be easily manufactured, stored, and administered.

James Baker, from the University of Michigan, presented his work in developing nanoemulsions as vehicles for vaccine and drug delivery. Nanoemulsions consist of nanometer size droplets of vegetable oil emulsified in water and stabilized with surfactants. Nanoemulsions function in a manner similar to detergents, disrupting microorganisms, but with much broader antimicrobial activity than detergents, killing not just bacteria, enveloped viruses, and fungi, but also bacterial and fungal spores and protozoa.

Despite good tissue penetrance, these emulsions have better toxicity profiles than detergents because they are less efficient at disrupting tissue. This property led to their evaluation as a treatment/prophylaxis for anthrax exposure, with Phase II human toxicity trials currently underway.

Investigators in Dr. Baker's lab demonstrated a protective immune response against vaccinia virus, HIV GP120/160, hepatitis-B surface antigen, and anthrax protective antigen. Experiments with anthrax protective antigen (PA) also showed that the addition of synthetic CpG oligonucleotides to the nanoemulsion protein mix could improve immune responses, so the system can potentially incorporate other adjuvants. Importantly, these results are obtained without the use of the cholera toxin, lipopolysaccharide, or bacterial proteins that are the mainstays of other mucosal adjuvants.

Another novel approach to using nanotechnology for therapeutics was presented by Naomi Halas, from Rice University. Nanoshells, developed by Dr. Halas's lab, are an example of a biocompatible nanostructure whose geometry can be modified to manipulate light in controlled ways. By adjusting the relative core and shell thickness, nanoshells can be manufactured to absorb or scatter light at a desired wavelength across visible and NIR wavelengths. Applications of these nanoshells include rapid immunoassays and optically addressable drug delivery materials. Dr. Halas also discussed her targeted cancer therapy research that is currently underway in mice.

In this study, nanoshell therapy was used to treat tumors in mice. Polyethylene glycol coated nanoshells with peak optical absorption in the NIR were intravenously injected into the mice and allowed to circulate for six hours. Palpable tumors were then illuminated with a diode laser for three minutes, heating the localized nanoshells and the tumor surface. The tumors were completely cleared with no recurrence ninety days post therapy without any side effects. The immunologists on the panel speculated as to whether the therapy was stimulating the host immune system to destroy the tumor or whether the therapy itself was responsible for destruction of the tumor.

MATERIALS SCIENCE

Construction of new nanomaterials and nanostructures are essential to the continued progress of nanotechnology into practical scientific and medical devices. Jeffrey Carbeck, from Princeton University, presented new strategies for fabrication and assembly. This work combines two existing methods, top down fabrication and bottom up assembly to pattern and orient single DNA molecules. These single DNA chains are then used as templates for the construction of hetero-structures composed of proteins and synthetic nanoparticles.

Template synthesized silica nanotubes can be potential candidates for the development of highly sensitive and selective biolabels. There are several unique properties that make nanotubes good candidates for this purpose: nanotubes can be filled with large amounts of fluorophores so that "colour-coding" can take place, the walls of nanotubes are completely transparent for UV visible light to pass through, nanotubes can be easily tailored to the right size for viewing through a traditional fluorescence microscope, and nanotubes have a high number of specific binding events along their axis as compared to spherically shaped particles.

Sang Bok Lee, from the University of Maryland, described his methods for loading high amounts of fluorescent dyes into the nanotubes by using hydrophobic interactions between organic and inorganic dyes and the inner surfaces of the nanotubes. He also discussed his experiments to optimize the binding affinity and specificity by lengthening the nanotubes, thus creating the potential to develop more sensitive and specific assays .

Although recent advances in chip technologies, such as microarrays and protein chips, have increased our ability to rapidly identify biomarkers and make diagnoses, there are many technical challenges to overcome to increase the sensitivity and selectivity of such uses. A significant limiting factor is nonspecific binding. In selective binding, one must modify a surface in order to block nonspecific interactions and at the same time promote specific binding.

Polyethylene glycol generally has been used as a tethering molecule to segregate the desired molecules from those not required for an experiment. The effects that various physical parameters have on binding are essentially unknown. Joyce Wong, from Boston University, presented her recent work on multiplexed screening of receptor-ligand binding. Her studies show that the presence of a tether significantly alters the range of receptor-ligand binding and is dependent on the dynamics of the tether. It is therefore important to be able to determine how these interactions affect the final outcome.

One of the promises of nanotechnology in biodefense or other biological applications is the potential for rapid detection and quantification of very small samples, such as those from single cells. Real time monitoring of the chemical output of the immune system is also a potential benefit. Harold Craighead, from Cornell University, discussed nanofabrication approaches to create devices for molecular detection, separation and quantification of small samples. These methods include the use of nanofluidic channels for single molecule counting and sizing, which could eventually be used to create real time immunoassays.

Shuichi Takayama, from the University of Michigan, discussed his vision for a platform for cell engineering and analysis that integrates computer controlled microfluidic cell culture and analysis with protein nanopatterned substrates that are tunable and can reversibly switch the behaviour of attached cells. Such a system would be useful for the study of complex interactions among various cellular and molecular components of the immune system.

BASIC IMMUNOLOGY

Developing treatments and vaccines in preparation for a bioterrorist attack will require a thorough understanding of how the immune system responds to different biological weapons. To help nanoengineers understand obstacles faced by the immunology community in dissecting immune mechanisms, several talks on basic immunology were presented.

During an immune response, many types of effector functions can be deployed against a pathogen. The choice of effector function is critical, as only certain mechanisms are effective against each pathogen, and unnecessary functions can cause severe host tissue damage. Cytokines secreted by T lymphocytes not only mediate some of the effector functions, but also regulate the overall type of immune response. Although considerable information has been obtained on the Th1 and Th2 T cell subtypes that secrete different sets of cytokines, it is clear that there is more T cell diversity than just these two types.

Studies of T cell populations are useful but limited because T cells behave asynchronously. Furthermore, the relevant T cells in an immune response are often present at low frequencies, e.g. less than 1 in 10,000. For all these reasons, single-cell assays for multiple T cell functions are valuable, particularly if the assays can monitor sequential events in individual cells.

Tim Mossman, from the University of Rochester, discussed his modification of the standard Elispot assay into a multiparameter Fluorispot assay using fluorescence detection. Three (and potentially more) cytokines can be detected simultaneously, allowing the evaluation of complex cytokine secretion phenotypes among low frequencies of antigen-specific cells. This assay has also been combined with the Lysispot assay, also developed by Dr. Mossman's lab, that detects individual cytotoxic T cells by measuring the release of a marker protein from target cells lysed by antigen-specific T cells.

T cells play a central role in the adaptive immune response to pathogens, by directly eliminating infected cells (cytotoxicity) and by promoting antibody and inflammatory responses (helper activity). The ability of T cells to recognize and respond to foreign materials is determined by the precise reactivity of their T cell antigen receptors (TCR) with antigen peptides complexed to major histocompatibility complex (MHC) proteins present on the surfaces of other cells.

Analysis of the antigen-specific T cell response is challenging, because of the low abundance of T cells responding to a particular pathogen and the relatively weak interaction of T cell receptors with their specific antigen-MHC complexes. Lawrence Stern, from the University of Massachusetts Medical School, discussed recent advances in the development of synthetic assemblies of recombinant MHC proteins in complex with defined peptide antigens. These assemblies have proven useful in the detection and analysis of antigen-specific T cells.

Macrophages represent one of the cornerstones of the innate immune system. They detect infectious organisms via a plethora of receptors; they phagocytose them, and then orchestrate an appropriate host response to them. In order to precisely define the nature of the threat, the macrophage needs to read the molecular bar code that is displayed on the specific pathogen. Recent work from a number of laboratories indicates that the family of Toll-like receptors (TLRs) plays a key role in defining the invading microorganism. Thus activation of TLR4, that detects LPS, and TLR5, that detects flagellin, would indicate the presence of a Gram negative, flagellated bacterium. This precise recognition triggers a highly regulated response to the pathogen by the host.

Alan Aderem, from the Institute for Systems Biology (ISB), described how genomic and proteomic tools developed at ISB define branch points in the TLR signaling pathways. He also discussed the need for sophisticated computational methods for analyzing the increasing amounts of complex data that will be generated by nanobiology. Dr. Aderem's team has begun to address this need by creating computational methods for defining regulatory points in the TLR signaling pathways.

Respiratory viral infections induce populations of peripheral memory T cells that persist in the pleural cavity, lung parenchyma and lung airways for the life of the individual. While relatively low in number, these cells are able to mediate substantial control of secondary challenge by engaging the virus at the site of infection and when viral loads are low. Understanding the factors that regulate the recruitment and maintenance of memory T cells in different lung compartments is essential for the development of vaccines designed to promote effective cellular immunity in the respiratory tract.

To better understand the regulation of T cell memory, David Woodland, from the Trudeau Institute, has been investigating the factors regulating the trafficking and maintenance of memory T cell populations in the lung. Current data suggest a dynamic process of recruitment and loss of cells with complete replacement of certain populations occurring within days.

However, little is known about which subpopulations of memory T cells are specifically recruited, how they traffic into distinct compartments, or what signals are required for recruitment. Dr. Woodland outlined some technological problems associated with analysis of the very small numbers of cells needed for his research. There is a need for highly specific and sensitive methods to track low cell numbers, a need to develop methods to study trafficking in situ, and a need to develop methods to apply these techniques in vivo or in live tissue.

A recurrent theme throughout the meeting was the need for development of new types of reagents that could replace antibodies. Many nanotechnologies in biomedical applications rely on antibodies for target identification. This can pose significant limitations on sensitivity and specificity in addition to problems with cross reactivity. Creation of a standard body or repository for current approved reagents and support for the development of new types of reagents specifically for nanobiology applications are needed.

In communication subsequent to the meeting, Dr. Takayama also suggested that, in addition to reagents, there is a need for support for the development of devices and materials specific to nanobiology applications and the creation of a repository of such devices and materials that could be made available to researchers.

Chapter 5

DNA Microarray

A DNA microarray is a multiplex technology used in molecular biology and in medicine. It consists of an arrayed series of thousands of microscopic spots of DNA oligonucleotides, called features, each containing picomoles of a specific DNA sequence. This can be a short section of a gene or other DNA element that are used as probes to hybridize a cDNA or cRNA sample (called target) under high-stringency conditions. Probe-target hybridization is usually detected and quantified by fluorescence-based detection of fluorophore-labeled targets to determine relative abundance of nucleic acid sequences in the target.

In standard microarrays, the probes are attached to a solid surface by a covalent bond to a chemical matrix (via epoxy-silane, amino-silane, lysine, polyacrylamide or others). The solid surface can be glass or a silicon chip, in which case they are commonly known as *gene chip* or colloquially *Affy chip* when an Affymetrix chip is used.

Other microarray platforms, such as Illumina, use microscopic beads, instead of the large solid support. DNA arrays are different from other types of microarray only in that they either measure DNA or use DNA as part of its detection system. DNA microarrays can be used to measure changes in expression levels, to detect single nucleotide polymorphisms (SNPs), in genotyping or in resquencing mutant genomes. Microarrays also differ in fabrication, workings, accuracy, efficiency, and cost. Additional factors for microarray experiments are the experimental design and the methods of analyzing the data.

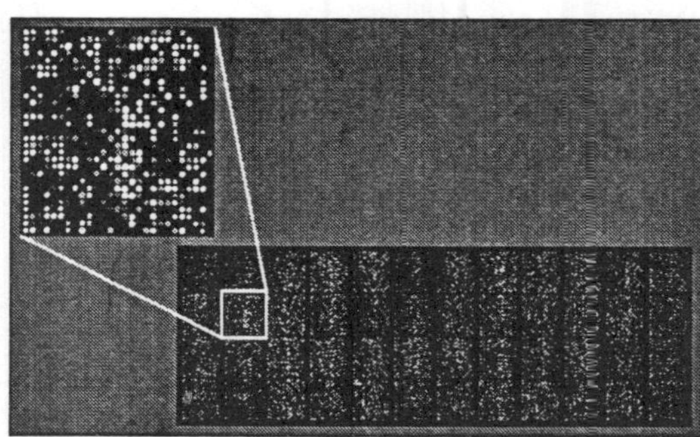

Fig. Example of an Approximately 40,000 Probe Spotted Oligo Microarray with Enlarged Inset to Show Detail

Microarray technology evolved from Southern blotting, where fragmented DNA is attached to a substrate and then probed with a known gene or fragment.

The use of a collection of distinct DNAs in arrays for expression profiling was first described in 1987, and the arrayed DNAs were used to identify genes whose expression is modulated by interferon.

These early gene arrays were made by spotting cDNAs onto filter paper with a pin-spotting device. The use of miniaturized microarrays for gene expression profiling was first reported in 1995, and a complete eukaryotic genome (*Saccharomyces cerevisiae*) on a microarray was published in 1997.

USES AND TYPES

Arrays of DNA can be spatially arranged, as in the commonly known *gene chip* (also called *genome chip, DNA chip* or *gene array*), or can be specific DNA sequences labelled such that they can be independently identified in solution. The traditional solid-phase array is a collection of microscopic DNA spots attached to a solid surface, such as glass, plastic or silicon biochip.

The affixed DNA segments are known as *probes* (although some sources use different terms such as *reporters*). Thousands of them can be placed in known locations on a single DNA microarray.

Technology or Application	*Synopsis*
Gene expression profiling	In an mRNA or gene expression profiling experiment the expression levels of thousands of genes are simultaneously monitored to study the effects of certain treatments, diseases, and developmental stages on gene expression. For example, microarray-based gene expression profiling can be used to identify genes whose expression is changed in response to pathogens or other organisms by comparing gene expression in infected to that in uninfected cells or tissues.
Comparative genomic hybridization	Assessing genome content in different cells or closely related organisms.
Chromatin immunoprecipitation on Chip	DNA sequences bound to a particular protein can be isolated by immunoprecipitating that protein (ChIP), these fragments can be then hybridized to a microarray (such as a tiling array) allowing the determination of protein binding site occupancy throughout the genome. Example protein to immunoprecipitate are histone modifications, Polycomb-group protein and trithorax-group protein to study the epigenetic landscape or RNA Polymerase II to study the transcription lanscape.
SNP detection	Identifying single nucleotide polymorphism among alleles within or between populations. Several applications of microarrays make use of SNP detection, including Genotyping, forensic analysis, measuring predisposition to disease, identifying drug-candidates, evaluating germline mutations in individuals or somatic mutations in cancers, assessing loss of heterozygosity, or genetic linkage analysis.
Alternative splicing detection	An exon junction array design uses probes specific to the expected or potential splice sites of predicted exons for a gene. It is of intermediate density, or coverage, to a typical gene expression array (with 1-3 probes per gene) and a genomic tiling array (with hundreds or thousands of probes per gene). It is used to assay the expression of alternative splice forms of a gene. Exon arrays have a different design, employing probes designed to detect each individual exon for known or predicted genes, and can be used for detecting different splicing isoforms.

DNA microarrays can be used to detect DNA (as in comparative genomic hybridization), or detect RNA (most commonly as cDNA after reverse transcription) that may or may not be translated into proteins. The process of measuring gene expression via cDNA is called expression analysis or expression profiling.

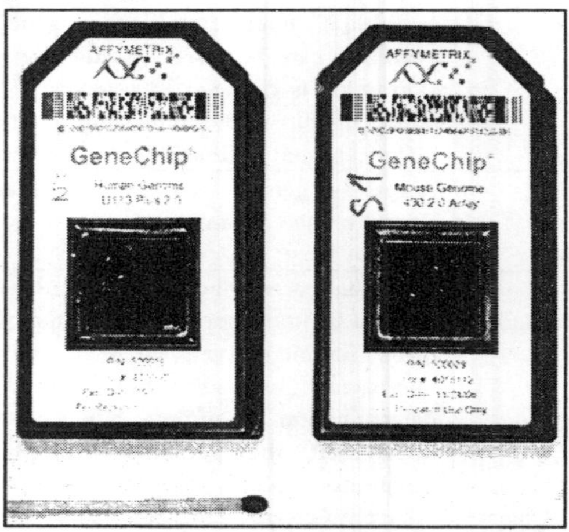

Fig. Two Affymetrix Chips

Since an array can contain tens of thousands of probes, a microarray experiment can accomplish that many genetic tests in parallel. Therefore arrays have dramatically accelerated many types of investigation.

Fabrication

Microarrays can be manufactured in different ways, depending on the number of probes under examination, costs, customization requirements, and the type of scientific question being asked. Arrays may have as few as 10 probes to up to 2.1 million (NimbleGen, Roche) micrometre-scale probes from commercial vendors.

Surface Engineering

The first step of DNA microarray fabrication involves surface engineering of a substrate in order to obtain desirable surface properties for the application of interest. Optimal surface properties are those which produce high signal to noise ratios for the DNA targets of interest. Generally, this involves maximizing the probe surface density and activity while minimizing the non-specific binding of the targets of interest. Methods of surface engineering vary depending on the platform material, design, and application.

Spotted vs. Oligonucleotide Arrays

Microarrays can be fabricated using a variety of technologies, including printing with fine-pointed pins onto glass slides, photolithography using pre-made masks, photolithography using dynamic micromirror devices, ink-jet printing, or electrochemistry on microelectrode arrays.

In *spotted microarrays*, the probes are oligonucleotides, cDNA or small fragments of PCR products that correspond to mRNAs. The probes are synthesized prior to deposition on the array surface and are then "spotted" onto glass. A common approach utilizes an array of fine pins or needles controlled by a robotic arm that is dipped into wells containing DNA probes and then depositing each probe at designated locations on the array surface.

The resulting "grid" of probes represents the nucleic acid profiles of the prepared probes and is ready to receive complementary cDNA or cRNA "targets" derived from experimental or clinical samples. This technique is used by research scientists around the world to produce "in-house" printed microarrays from their own labs.

These arrays may be easily customized for each experiment, because researchers can choose the probes and printing locations on the arrays, synthesize the probes in their own lab (or collaborating facility), and spot the arrays. They can then generate their own labeled samples for hybridization, hybridize the samples to the array, and finally scan the arrays with their own equipment.

This provides a relatively low-cost microarray that may be customized for each study, and avoids the costs of purchasing often more expensive commercial arrays that may represent vast numbers of genes that are not of interest to the investigator. Publications exist which indicate in-house spotted microarrays may not provide the same level of sensitivity compared to commercial oligonucleotide arrays, possibly owing to the small batch sizes and reduced printing efficiencies when compared to industrial manufactures of oligo arrays.

In *oligonucleotide microarrays*, the probes are short sequences designed to match parts of the sequence of known or predicted open reading frames. Although oligonucleotide probes are often used in "spotted" microarrays, the term "oligonucleotide array" most often refers to a specific technique of manufacturing. Oligonucleotide arrays are produced by printing short oligonucleotide sequences designed to represent a single gene or family of gene splice-variants by synthesizing this sequence directly onto the array surface instead of depositing intact sequences.

Sequences may be longer (60-mer probes such as the Agilent design) or shorter (25-mer probes produced by Affymetrix) depending on the desired purpose, longer probes are more specific to individual target genes, shorter probes may be spotted in higher density across the array and are cheaper to manufacture.

One technique used to produce oligonucleotide arrays include photolithographic synthesis (Agilent and Affymetrix) on a silica substrate where light and light-sensitive masking agents are used to "build" a sequence one nucleotide at a time across the entire array. Each applicable probe is selectively "unmasked" prior to bathing the array in a solution of a single nucleotide, then a masking reaction takes place and the next set of probes are unmasked in preparation for a different nucleotide exposure.

After many repetitions, the sequences of every probe become fully constructed. More recently, Maskless Array Synthesis from NimbleGen Systems has combined flexibility with large numbers of probes.

Two-channel vs. One-channel Detection

Two-colour microarrays or *two-channel microarrays* are

typically hybridized with cDNA prepared from two samples to be compared (e.g. diseased tissue versus healthy tissue) and that are labeled with two different fluorophores. Fluorescent dyes commonly used for cDNA labelling include Cy3, which has a fluorescence emission wavelength of 570 nm (corresponding to the green part of the light spectrum), and Cy5 with a fluorescence emission wavelength of 670 nm (corresponding to the red part of the light spectrum).

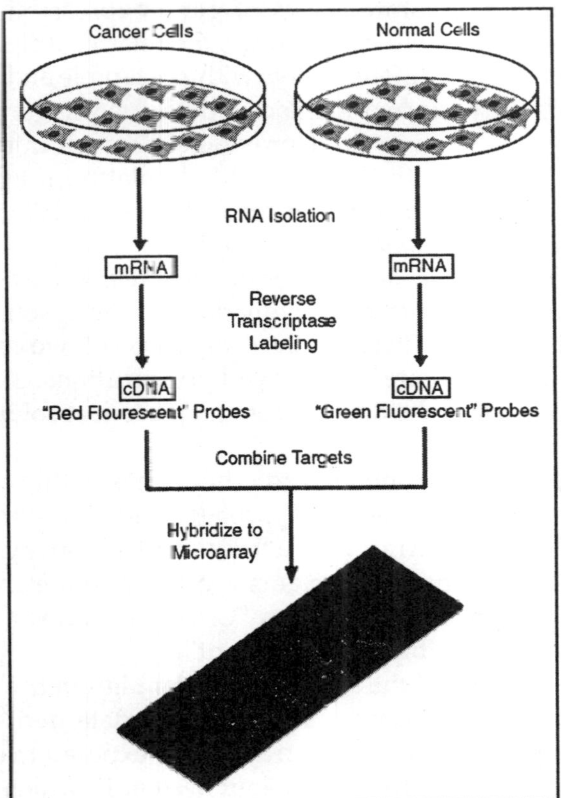

Fig. Diagram of Typical Dual-colour Microarray Experiment

The two Cy-labelled cDNA samples are mixed and hybridized to a single microarray that is then scanned in a microarray scanner to visualize fluorescence of the two fluorophores after excitation with a laser beam of a defined wavelength. Relative intensities of each fluorophore may then be used in ratio-based analysis to identify up-regulated and down-regulated genes.

Oligonucleotide microarrays often contain control probes designed to hybridize with RNA spike-ins. The degree of hybridization between the spike-ins and the control probes is used to normalize the hybridization measurements for the target probes.

Although absolute levels of gene expression may be determined in the two-colour array, the relative differences in expression among different spots within a sample and between samples is the preferred method of data analysis for the two-colour system. Examples of providers for such microarrays includes Agilent with their Dual-Mode platform, Eppendorf with their DualChip platform for fluorescence labeling, and TeleChem International with Arrayit.

In *single-channel microarrays* or *one-colour microarrays*, the arrays are designed to give estimations of the absolute levels of gene expression. Therefore the comparison of two conditions requires two separate single-dye hybridizations. As only a single dye is used, the data collected represent absolute values of gene expression.

These may be compared to other genes within a sample or to reference "normalizing" probes used to calibrate data across the entire array and across multiple arrays. Three popular single-channel systems are the Affymetrix "Gene Chip", the Applied Microarrays "CodeLink" arrays, and the Eppendorf "DualChip & Silverquant".

One strength of the single-dye system lies in the fact that an aberrant sample cannot affect the raw data derived from other samples, because each array chip is exposed to only one sample (as opposed to a two-colour system in which a single low-quality sample may drastically impinge on overall data precision even if the other sample was of high quality).

Another benefit is that data are more easily compared to arrays from different experiments; the absolute values of gene expression may be compared between studies conducted months or years apart. A drawback to the one-colour system is that, when compared to the two-colour system, twice as many microarrays are needed to compare samples within an experiment.

Microarrays and Bioinformatics

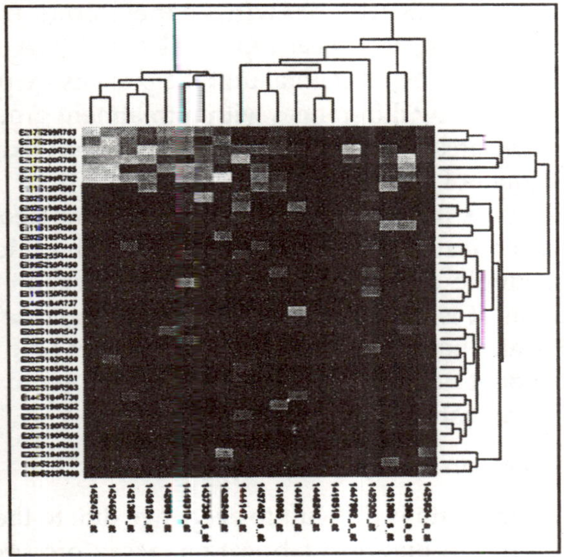

Fig. Gene Expression Values from Microarray Experiments can be Represented as Heat Maps to Visualize the Result of Data Analysis

The advent of inexpensive microarray experiments created several specific bioinformatics challenges:

- The multiple levels of replication in experimental design (Experimental design)
- The number of platforms and independent groups and data format (Standardization)
- The treatment of the data (Statistical analysis)
- What exactly are we measuring (Relation between probe and gene)

Experimental Design

Due to the biological complexity of gene expression, the considerations of experimental design that are discussed in the expression profiling article are of critical importance if statistically and biologically valid conclusions are to be drawn from the data.

There are three main elements to consider when designing a microarray experiment. First, replication of the biological samples is essential for drawing conclusions from the experiment. Second, technical replicates (two RNA samples obtained from each experimental unit) help to ensure precision and allow for testing differences within treatment groups. The technical replicates may be two independent RNA extractions or two aliquots of the same extraction.

Third, spots of each cDNA clone or oligonucleotide are present as replicates (at least duplicates) on the microarray slide, to provide a measure of technical precision in each hybridization. It is critical that information about the sample preparation and handling is discussed, in order to help identify the independent units in the experiment and to avoid inflated estimates of statistical significance.

Standardization

Microarray data is difficult to exchange due to the lack of standardization in platform fabrication, assay protocols, and analysis methods. This presents an interoperability problem in bioinformatics. Various grass-roots open-source projects are trying to ease the exchange and analysis of data produced with non-proprietary chips:

- For example, the "Minimum Information About a Microarray Experiment" (MIAME) checklist helps define the level of detail that should exist and is being adopted by many journals as a requirement for the submission of papers incorporating microarray results. But MIAME does not describe the format for the information, so while many formats can support the MIAME requirements, as of 2007 no format permits verification of complete semantic compliance.
- The "MicroArray Quality Control (MAQC) Project" is being conducted by the US Food and Drug Administration (FDA) to develop standards and

quality control metrics which will eventually allow the use of MicroArray data in drug discovery, clinical practice and regulatory decision-making.

- The MGED Society has developed standards for the representation of gene expression experiment results and relevant annotations.

Statistical Analysis

The analysis of DNA microarrays poses a large number of statistical problems, including the normalization of the data. There are several normalization methods in the published literature some of which are platform specific; as in many other cases where authorities disagree, a sound conservative approach is to directly compare different normalization methods to determine the effects of these different methods on the results obtained. This can be done, for example, by investigating the performance of various methods on data from "spike-in" experiments.

Also, experimenters must account for multiple comparisons: even if the statistical P-value assigned to a gene indicates that it is extremely unlikely that differential expression of this gene was due to random rather than treatment effects, the very high number of genes on an array makes it likely that differential expression of some genes represent false positives or false negatives.

Statistical methods tailored to microarray analyses have recently become available that assess statistical power based on the variation present in the data and the number of experimental replicates, and can help minimize type I and type II errors in the analyses.

A basic difference between microarray data analysis and much traditional biomedical research is the dimensionality of the data. A large clinical study might collect 100 data items per patient for thousands of patients. A medium-size microarray study will obtain many thousands of numbers per sample for perhaps a hundred samples.

Many analysis techniques treat each sample as a single point in a space with thousands of dimensions, then attempt by various techniques to reduce the dimensionality of the data to something humans can visualize.

Relation between Probe and Gene

The relation between a probe and the mRNA that it is expected to detect is problematic. On the one hand, some mRNAs may cross-hybridize probes in the array that are supposed to detect another mRNA. On the other hand, probes that are designed to detect the mRNA of a particular gene may be relying on genomic EST information that is incorrectly associated with that gene.

Microarray data was found to be more useful when compared to other similar datasets. The sheer volume (in bytes), specialized formats (such as MIAME), and curation efforts associated with the datasets require specialized databases to store the data.

Chapter 6

Hemicelluloloses in the Nanobiology

The hemicelluloses have not received adequate attention in studies of wood cell walls because the complexity of their structures does not admit easy interpretation within the paradigms of polymer science. Two-phase composite models of the cell wall have led many to view their primary function as one of coupling cellulose and lignin to enhance the mechanical properties of the walls. But that is a microscopic interpretation based on macroscopic theories of reinforced structures. In contrast, recent studies have shown that hemicelluloses can participate in regulation of the nanoscale architecture of cell wall constituents.

They influence the aggregation of celluloses, and they can also influence the pattern of inter-unit linkages in lignin analogs polymerized in their presence. It is clear that the hemicelluloses have higher functions than as mechanical coupling agents in a two-phase composite. The hemicelluloses are species and tissue specific, and they have also been shown to vary across the layers of cell walls in patterns that are similar for cells formed contemporaneously in the same tissue. Their structures are clearly genetically encoded. In this report it is proposed that the hemicelluloses have multiple functions in regulation of cell wall consolidation and in determining its properties.

The hemicelluloses are viewed as part of a system whereby information encoded in the genes is communicated to regulate the assembly of the cell wall. In addition to their role in regulating the aggregation of cellulose and the formation of inter-unit linkages in lignin, they appear to act in concert with extra-cellular glycosidases, to accomplish gradual dehydration of the polysaccharide matrix within which lignin precursors are polymerized.

These proposals are based on viewing the deposition and consolidation of the secondary wall as the result of multiple, simultaneous and sequential processes for synthesis of cell wall constituents wherein the syntheses are distributed in both space and time. The spatially distributed syntheses are considered essential to allow gradual modification of the molecules of the precursors of cell wall constituents. They begin as molecules optimized for solubility within the aqueous environment in the cell and are transformed into ones that can be consolidated within the secondary wall, which is more hydrophobic upon lignification.

To add clarity to the analysis, a system theoretic approach to the discussion of the distribution of the biogenetic processes in space is adopted. The overall system within each cell is regarded as three highly coupled subsystems. The nanoscale organization in the native state is then examined as a reflection of the processes of biogenesis of the plant cell walls. Issues associated with aggregation of the constituents are then considered and plausible pathways for the assembly of the matrix that are based on distributed synthesis of the major polymeric constituents are suggested.

The molecular organization is regarded as the first stationary level of expression of phenotypic form and the implications of the similarity of this expression within contemporaneous cells in an annual ring of secondary growth are examined. It is concluded that all aspects of molecular organization must be governed by intracellular processes that can be orchestrated at levels well beyond those of the individual cells.

It is proposed that the primary molecular carriers of organizing information, that is the mediators of the orchestration, are the hemicelluloses, and the manner of their action in the formation of the structure of lignin is suggested.

The results of the analysis of hierarchic organization have a number of additional implications with respect to structure and its formation that are beyond the scope of the present report. One that is worthy of note at this time is that the organization of the deposition of the secondary wall, and its systematic variation within different tissues in higher plants, requires intimate coupling and orchestration of intracellular, membrane and extracellular processes, at levels higher than that of the individual cell in a manner that has not heretofore been considered.

The hemicelluloses have become the orphan constituents of wood cell walls. They have been neglected for many years. One of the reasons is that the complexity of their structures does not admit easy interpretation within the paradigms of polymer science. These paradigms were originally developed in the context of investigations of the natural homo polymers cellulose, starch and natural rubber. However, since the 1950s further development of the paradigms for macromolecules has been based primarily on studies of synthetic polymers.

Within these newer conceptual frameworks, the hemicelluloses have frequently been classified as random copolymers. The two-phase composite model of the cell wall has led many to view the primary function of the hemicelluloses as one of coupling celluloses and lignin to enhance the mechanical properties of the composite structures of cell walls. But that is a microscopic level interpretation growing from macroscopic theories of reinforced structures and theories of fibre-matrix composites.

In contrast, recent studies have shown that the hemicelluloses can participate in regulation of the nanoscale architecture of cell wall constituents. They influence the aggregation of celluloses, and they can also influence the pattern of inter-unit linkages in lignin analogs polymerized in their presence. It is plausible therefore to consider the possibility that the hemicelluloses have higher functions than as mechanical coupling agents in a two-phase composite.

It seems unlikely that evolutionary adaptation would have produced hemicelluloses as diverse, and as species and tissue specific as those that have been identified so far, for such a relatively simple function. Not only are the hemicelluloses species and tissue specific, but they have also been shown to vary across the layers of cell walls. Their structures are clearly genetically encoded.

In this report it is proposed that the hemicelluloses have multiple functions in regulation of cell wall consolidation and in determining its properties. It will be suggested that the hemicelluloses are part of a system whereby information encoded in the genes is communicated to regulate the assembly of the cell wall. In addition to their role in regulating the aggregation of cellulose and the formation of inter-unit linkages in lignin, they appear to act in concert with extra-cellular glycosidases, to accomplish gradual dehydration of the polysaccharide matrix within which lignin precursors are polymerized.

These proposals are based on viewing the deposition and consolidation of the secondary wall as the result of multiple, simultaneous and sequential processes for synthesis of cell wall constituents wherein the syntheses are distributed in both space and time. The spatially distributed syntheses are considered essential to allow gradual modification of the molecules of the precursors of cell wall constituents from molecules optimized for solubility within the fluid environment in the cell into ones that are eventually consolidated within the secondary cell wall that is essentially hydrophobic upon lignification.

To add clarity to the discussion, a system theoretic approach to the discussion of the distribution of the biogenetic processes in space will be adopted. The overall system within each cell will be regarded as three highly coupled subsystems. The first subsystem is the intracellular environment A wherein the precursors to cell wall constituents are synthesized in a hydrophilic form suitable for a predominantly aqueous environment.

This first subsystem is separated from the second one by the plasma membrane as envisioned in current plant cell theory. While multiple processes occur in the plasma membrane, for our purposes it is regarded as a boundary or interface between the intracellular A and the immediate extra cellular B environments or subsystems, and it is a boundary across which there is much molecular traffic as the biogenesis unfolds.

The second subsystem B is regarded as the highly hydrated extra cellular layer immediately adjacent to the membrane outside the cell during cell wall formation. It will be suggested that this subsystem is the context of additional modifications of the molecules that have emerged from the plasma membrane to prepare them for consolidation into the cell wall.

The third subsystem C is regarded as the consolidating and lignifiying cell wall bounded by the primary wall and cell corners D on the outside and the less well defined interface between it and subsystem B on the inside. In the normal course of events the boundaries will move as the cell wall is deposited and the second subsystem B will disappear when the cell wall is fully formed.

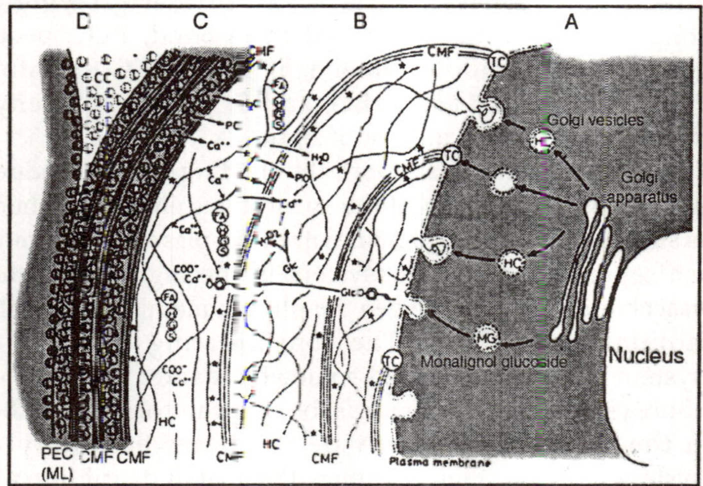

Fig. Conceptual Illustration of Biogenesis of Cell Wall. Golgi Apparatus Participates in Biosynthesis of Cellulose, Hemicellulose, and Lignin. Cellulose Microfibrils are Laid Separately in Swollen gel of hemicellulose

We note here that this perspective has evolved in part from our collaborations with Professor Emeritus Noritsugu Terashima of Nagoya University and his extensive studies with his co-workers, focused on the processes of lignification during cell wall biogenesis. Many discussions with him have influenced our views of biogenesis.

Two of the figures used herein are adaptations of figures that he has used to illustrate his views concerning the time sequence of deposition of lignin precursors and of lignification. However we assume responsibility for the adaptations and interpretations we use to complement those of Professor Terashima.

Lignin deposition proceeds from intercellular region to inner part of the cell wall, and becomes hydrophobic. Water is removed from the swollen gel together with peroxidase (PO) and Ca toward inner side of the wall to cause an anisotropic shrinkage in the direction perpendicular to cellulose microfibril (CMF).

This shrinkage brings about orientation of aromatic ring parallel to the cell wall surface and further growing of oligolignol to high polymer: hydrogen bond, PEC: pectic substances, ML: middle lamella, L: lignin, TC: terminal complex, FA: ferulic acid, H,G, and S: p-hydroxyphenyl, guaiacyl, and syringyl moiety of lignin.

In adaptation for this report, the shading has been added to indicate the boundaries of the subsystems envisioned here. Thus subsystem A is the domain within the plasma membrane, which separates it from subsystem B, the highly hydrated polysaccharide matrix between the plasma membrane and the consolidating cell wall. The consolidating cell wall is subsystem C which is shaded somewhat darker than A.

Subsystem D defined here as the cell wall corner, together with the primary wall provide the outer boundary of subsystem C. While Figure depicts the spatial distribution of processes, Figure, also developed by Professor Terashima and coworkers indicates the temporal sequence in the formation of the cell wall.

Cell wall layers are formed successively in the order of

middle lamella (ML/cell corner (CC), primary wall (PW), followed by the outer, middle, and inner layers of the secondary walls (SW)(S1, S2 and S3 layers). Monolignol units are incorporated in the order of p-hydroxyphenyl (H), guaiacyl (G), and syringyl (S) units, and incorporated thus in three distinct stages.

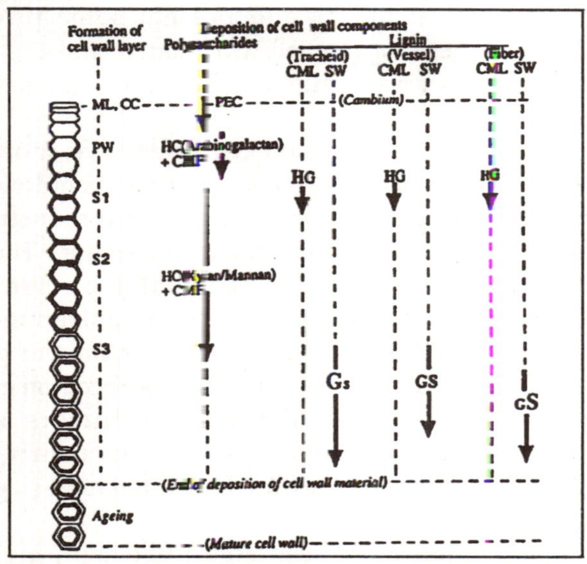

Fig. A conceptual Illustration of the Relationship between Formation of Cell Wall Layers and Incorporation of Polysaccharide and Lignin

The first lignification takes place in the CC and compound middle lamella (CML) after the PW is formed and just before S1 formation starts. Thereafter slow (or no) lignification takes place during the deposition of cellulose microfibrils (CMFs) and hemicelluloses (HC) in S2. The most extensive deposition of lignin takes place in the S2, after S3 formation has started.

In order to establish the perspective that has informed our discussion, we begin with an overview of our findings concerning molecular organization in secondary walls of woody tissue. We then provide a brief account of prevailing views concerning biosynthesis of major cell wall polymers; this is necessarily brief and limited in scope. We next examine issues of aggregation and distributed synthesis. We follow with a discussion of relationships between adjacent cells, the degree to which processes of biogenesis in neighboring cells are coupled, and the extension of the coupling to cells further removed within an annual ring of secondary growth. Finally, we consider the combined implications of information at different levels of structure and present some proposals concerning biogenesis of the cell wall matrix and the role of the hemicelluloses and cellulose as primary factors in its organization.

Studies of biogenesis of plant cell walls have, most often, addressed biosynthesis of the major constituents individually. In general the investigative approach has been reductive and there have been only limited efforts at integration. The focus in studies of polysaccharides is usually on synthases responsible for their formation, and on organization of the synathases within the Golgi apparatus or on the plasma membrane. In studies of lignin, the focus has been on enzyme systems responsible for synthesis of precursors and on processes of polymerization in the extracellular matrix. Issues of aggregation and tertiary structure formation have in general received less attention.

Our own studies have been concerned with molecular organization in wood cell walls where aggregation and consolidation of the cell wall matrix become paramount, particularly in secondary walls. Our findings required us to examine development of tertiary structure. We found that questions of tertiary structure cannot be separated from questions of assembly of the matrix, and that, in turn, assembly of the matrix cannot be separated from questions of biosynthesis of the constituents.

We have thus been motivated to seek a unifying model for biogenesis and assembly of the major constituents of cell walls. In our search, the role of the hemicelluloses was given special attention. Though this first grew out of the differences between wood secondary wall hemicelluloses and those that have been investigated in primary walls of non-woody plants, it eventually led us to a system theoretic approach to understanding the processes of cell wall formation. The purpose of this report is to present such a model which evolved from our contemplation of the factors that must be considered from a system theoretic perspective.

In our search for a unifying model, we have been guided by two classes of considerations associated with spatial organization of cellular systems. The first class includes those arising regarding aggregation of the constituents and of their precursors in the different solvating environments represented by the cytosol, the extracellular matrix and the consolidating walls. These led us to examine the possibility that the processes of biogenesis require distributed synthesis, that is, a sequence of synthetic steps that are distributed in space as well as separated in time. Hence the three subsystems proposed above, each representing a different environment wherein different categories of reactions can occur.

The second class of considerations are those flowing from recognizing that the molecular architecture of cell walls represents the first level of stationary expression of phenotypic form, and from examining the degree to which both the architecture and the assembly of the individual cell reflect its similarity to other cells within a particular tissue type. The symmetry of some structures at higher levels of organization, as in annual rings, for example, suggests a synchrony of molecular processes removed from each other in space, though not in time.

The considerations outlined above led us to explore the system and information theoretic perspectives set forth in the Introduction. We would emphasize that the intent of our proposals is not to negate the import of prior efforts in this arena, but rather to complement and integrate them on the basis of information derivable from examination of secondary wall structures and the potential role of the hemicelluloses. Where our proposals depart from previous models, we intend them as hypotheses that can rationalize observations, expecting that they will lead to new directions in research concerning the hemicelluloses, the results of which will allow discrimination between alternatives. We favor our own proposals primarily because we find them more plausible from both physical chemical and system theoretic perspectives.

We do not minimize the importance of cell wall proteins by excluding them from discussion of wall structure, for they are the primary agents of change in the synthetic processes in all subsystems. Rather, we focus primarily on the major constituents of the consolidated cell walls because they are the primary determinants of properties and because their native structures reflect the processes of assembly. They become in a sense terminal boundary conditions for the processes of assembly. The key role of the proteins in the dynamic geometry of the primary wall during morphogenesis is beyond the scope of this report.

TERTIARY STRUCTURE IN SECONDARY WALLS

The primary objective of our work has been to understand structure in the native state, that is, the organization of constituents of secondary walls prior to disruption by the relatively harsh processes needed for isolation of individual constituents. More recently, we recognized that the structures of the individual polymers of the cell wall are interdependent. It became necessary to integrate knowledge of the assembly of cellulose with knowledge of the formation and deposition of the hemicelluloses, and to reconcile our understanding of the polymerization of lignin with our models of the polysaccharide matrix where the polymerization occurs.

Our focus on the structure of secondary walls in woody tissues is motivated by their dominance in such tissue and by the fact that their deposition is not complicated by changes in geometry. A deeper understanding of formation of the secondary wall can contribute information complementary to that derived from studies of primary walls.

We regard primary walls as more complex due to their dynamic geometry, their thinness, and the high degree to which proteins participate in their development. Within secondary walls, our concern is with molecular organization at the nanoscale level, that is, domains of the order of 1 to 10 nm. This focus is motivated by the observation that collective properties of oligomers of cell wall constituents converge to those of the polymers within this size range.

Structures that give rise to collective properties of the intimately integrated constituents are well defined by organization at this level; the spectral signatures of oligomers of homopolymeric polysaccharides typically converge between the hexamer and the decamer, and for hemicelluloses, branching patterns are well defined within a backbone interval of 10 to 15 units.

OBSERVATIONS IN SUMMARY

When we commenced our work, the accepted models of cell wall structure were those described by of Preston and Frey-Wyssling. The cell wall was regarded as two phases, the cellulose fibrils, considered crystalline and homogeneous, and a surrounding matrix, seen as a poorly defined blend of hemicelluloses and lignin. Since then our early work showed that native cellulose is a composite of two distinct forms, Iá and Iâ, which occur in different proportions in different species; Iá is dominant in bacterial and algal celluloses while Iâ is dominant in celluloses from higher plants.

Later work revealed a variability in distribution of the hemicelluloses within wood cell walls. Pentosans are concentrated in the primary wall and the S1 and S3 layers, while hexosans are predominantly in the S2 layer, which is by far the largest fraction of matter in the walls. More recent work has shown that hemicelluloses are intimately blended into the structure of cellulose.

The hemicelluloses co-crystallize with cellulose, within elementary fibrils, to a degree that brings into question the validity of extending to higher plants a two-phase model of cell walls. That model may have been plausible for some of the primitive marine algae that were the focus of the early studies of cell wall celluloses, and which had nanofibrils that were approximately 20 nm in lateral dimensions, but it certainly cannot be regarded as adequate in systems wherein the elementary nanofibrils are of the order of 3 to 6 nm. The hemicelluloses also have been show capable of influencing the balance between Iá and Iâ in cell wall celluloses. Regarding the structure of lignin, our early work pointed to a higher degree of orientation than previously recognized and more recent results point to coupling between the organization of lignin and the structure of polysaccharides.

More specifically, early studies, using the Raman microprobe, demonstrated orientation of lignin relative to the plane of the cell wall and the cellulose within it, and also revealed variability in the ratio of lignin to cellulose in different locations within the wall. More recent results suggest that the cellulose and the hemicelluloses play an important role in the organization of lignin. The observation of photoconductivity in wood reflects a coherence of order in lignin that is sufficient to allow coupling between the lowest unoccupied molecular orbitals of the phenyl propane structural units.

That coherence of order can arise from association with a polysaccharide surface possessing some regularity has been demonstrated through molecular modeling studies. The influence of association between precursors and the matrix is also demonstrated in preparations of lignin analogs within cellulolse-hemicellulose matrices by dehydrogenative polymerization; the synthetic lignins are designated DHPs. The DHPs were found to be closer approximations to milled wood lignin (MWL) than any previously prepared. The implication of this finding and some earlier ones is that an association between the matrix and lignin precursors, strong enough to influence the pattern of interunit linkages, can play an important role in assembly of lignin.

The observations summarized, when considered together with earlier findings concerning variability of the hemicelluloses within secondary walls, suggested that hemicelluloses may be the key to a new model for assembly of lignin, one that could reconcile much of what is known about the polymerization of precursors with the requirements of intracellular control of overall assembly of the cell wall.

MODELS OF BIOSYNTHESIS

It is not within the scope of this report to provide an

extensive discussion of biosynthesis of constituents of the secondary wall. Perusal of the literature reveals dominance of distinctive paradigms regarding each of the major constituents. This is clear from examining some recent overview volumes dealing with both biogenesis and structure. The paradigms have considerable value in that they provide organizing principles for information in specific areas. When considering the cell wall as a whole, however, it is necessary to ask how well the processes can mesh together in an overarching model for assembly of cell walls.

While the complexity of the assembled walls has been recognized, the processes involved in the consolidation of the three major constituents have received limited attention. But the biosynthetic pathways for constituents must be integrated in the process of biogenesis of the walls, and the integration must be consistent with what is known of biological regulatory processes that govern phenotypic expression of genotypic information.

The Hemicelluloses

The area of greatest consensus is that of formation of hemicelluloses; it is generally agreed that they are assembled in final form within the Golgi apparatus and transported in vesicles for extracellular deposition by exocystosis. The processes involved have been reviewed in the context of wood formation, as well as in the broader context of plant cell wall development. A recent discussion of the function of the Golgi apparatus also addresses its role in the synthesis of complex cell wall polysaccharides.

Cellulose

The degree of consensus is more limited with respect to biogenesis of cellulose. There is evidence that membrane bound metabolically active entities are associated with assembly of cellulose microfibrils. Furthermore, cellulose synthase enzymes have been isolated from cellulose-forming bacteria, but the isolation of similar activity from higher plants remains elusive and controversial.

Different interpretations of the phenomenology of cellulose biogenesis have arisen. Some evidence supports a coupling of synthesis and crystallization in bacterial systems and for a number of eukaryotes. On the other hand, it is observed that there is a high degree of coordination between cytoskeletal architecture and patterns of cellulose deposition. Many studies have shown correlation between the organization of microtubules and the orientation of cellulose microfibrils. Also, treatments which disrupt organization of microtubules result in disruption of patterns of deposition of cellulose microfibrils. Thus, many questions remain concerning cellulose biogenesis.

Lignin

Least well defined is the process of assembly of lignin, where formation of precursors is intracellular and polymerization is extracellular. With the exception of the work by Terashima and coworkers cited above, research on biosynthesis of lignin has focused on formation of precursor monolignols. In addition, many studies have sought to understand lignification by investigating DHPs formed from monolignols under conditions that simulate different aspects of the microenvironment in the wall matrix.

Implicit in all studies of lignification is dominance of free radical coupling reactions described by Freudenberg. It is assumed that the distribution of linkage types is determined by the relative reaction rates between free radicals under conditions prevailing in the cell wall matrix; variations in the microenvironment are thought to influence electron density distributions within the radical intermediates.

ISSUES OF AGGREGATION AND THE NEED FOR DISTRIBUTED SYNTHESIS

The question of aggregation arises because consolidation of the cell wall matrix implies driving forces that result in association of the constituents simultaneously with progressive dehydration. The association is irreversible and the levels of hydration are well below those in subsystem B and the intracellular environment A. The aggregation cannot be regarded as a separation of phases; it is more akin to formation of insoluble aggregates in polymeric systems.

In the primary wall, consolidation reduces the degree of hydration but, prior to onset of lignification, the coupling between constituents remains weak enough that they can be separated. It is possible to separate pectic substances and hemicelluloses from the cellulose, although it is not always certain that the cellulose is free of small amounts of hemicelluloses.

Since research on cell wall matrix formation has focused on primary walls, issues of aggregation have required little attention. In secondary walls, in contrast, particularly in woody tissue, consolidation of the polysaccharide matrix involves a higher level of coupling of hemicelluloses with the cellulose. This is evident in holocelluloses prepared by mild oxidative delignification of woody tissue. The dominant hemicelluloses in secondary walls cannot be separated from cellulose without use of strong caustic solutions.

Yet all of these constituents have their origin in intracellular environments where they are highly solvated in aqueous media. It is necessary, therefore, to consider transformations that alter the relationships of the constituents to their environments. In particular, it is important to consider the likelihood that constituents that are inherently insoluble in water, or are of very limited solubility, are first synthesized in glycosylated forms to facilitate their solvation in the intracellular environment, and that they are later deglycosylated by action of one of the many glycosidases that occur in the extracellular environment in the vicinity of the plasma membrane.

The possibility that such transformations are central to assembly of the cell wall matrix arises with respect to all three major components of the cell wall. Such transformations can be regarded as a form of distributed synthesis, that is, a progression of synthetic steps that, in addition to being sequential in time, are distributed in space with respect to the domains within which they occur. Distributed synthesis is well established among intracellular processes; one instance is the synthesis of glycoproteins, which is thought to begin at the endoplasmic reticulum and to be completed within the Golgi apparatus.

Lignin Precursors

Distributed synthesis is most obvious with lignin precursors. These occur as monolignol glucosides within the cytoplasm and have been isolated by extraction of thin layers of differentiating xylem from a number of woody species. Significant quantities of coniferin and syringin have been isolated in this manner. As glucosides, these precursors of lignin are soluble in water to a level approaching 30%; upon separation of the glucose solubility declines to less than 1%.

It has been recognized for some time that extracellular glucosidases cleave the glycosidic linkages to liberate monolignols.

This has been confirmed by recent observation of a coniferin-specific glucosidase in lignifying tissues. Upon deglycosylation, monolignols can penetrate readily into the polysaccharide matrix which, when considered as a solvating medium, is similar to a mixture of water, alcohol and ether. Thus, the precursors of lignin represent an instance of a distributed synthetic process where the constituents are transformed along the synthetic pathway to alter their aggregative properties.

The Hemicelluloses

The principle of distributed synthesis also provides a basis for understanding the phenomenology of hemicelluloses, though the need for transformation is not obvious. In studies of primary walls the majority of constituents, with the exception of cellulose, can be isolated using mild extractive procedures. This possibility has led to a substantial literature on organization of the primary wall as well as its primary non-cellulosic polysaccharides, the xyloglucans and the pectins.

In secondary walls of woody tissues, in contrast, the

presence of major hemicellulose fractions would be difficult to comprehend apart from the possibility of distributed synthesis. The xylans in hardwoods and the glucomannans in softwoods include subsets that can be separated from cellulose only by extraction with strong caustic solutions. And they cannot be kept in solution upon dilution or neutralization; indeed progressive dilution and neutralization are the bases of methods for their fractionation.

If these hemicelluloses are also synthesized in the Golgi apparatus, it is very likely that they are assembled first in a soluble form, and are later modified prior to aggregation with cellulose.

The modification that suggests itself is analogous to that recognized for the precursors of lignin. If hemicelluloses are first synthesized with some limited substitution of mono- or disaccharide branches, they would be soluble in the intracellular environment. The action of extracellular glycosidases would then strip them of branches to facilitate aggregation with cellulose and with other â-1,4 linked glycans of limited substitution. Though occurrence of extracellular glycosidases has been recognized for some time, the function proposed here has not been considered among their roles. Thus, not unlike the lignin precursors, they originate in subsystem A, are modified in subsystem B and aggregate with cellulose in C.

An alternative rationalization of the problem of aggregation has been proposed by Bolwell. He suggested that hemicelluloses may be assembled in the extracellular environment from oligosaccharides. While this proposal addresses the issue of aggregation in the intracellular environment, it is not clear that such a process is consistent with the regularity characteristic of the structures of hemicelluloses.

This regularity suggests that the enzymes enabling assembly are organized in space relative to each other; such organization is implicit in the topology of the Golgi apparatus. If similar organizing structures occur in the extracellular matrix, they have not been reported. The kinetic implications and requirements of Bolwell's proposal have not been fully developed.

Cellulose

The possibility that a distributed synthetic process may be involved in the production of cellulose must also be considered. While such a proposal departs from accepted models, it is not at all inconsistent with much of the data available on cellulose synthesis, and it may provide a basis for rationalizing coupling of cytoskeletal organization with cellulose deposition. It could also explain the considerable difficulty encountered in efforts to isolate cellulose synthases from higher plants.

The process of distributed synthesis proposed for cellulose is similar to that proposed for hemicelluloses. Cellulose would be assembled first as a gluco-glucan, with a â-1,4 linked backbone and glucose branches, not unlike xyloglucan, but with glucose branches instead. Such a polymer would have escaped detection in studies that rely on hydrolysis and sugar analysis to validate the isolation of cellulose.

Colvin et al. isolated a gluco-glucan with a glucose substituent at the 2 position of every third backbone unit from cultures of Acetobacter xylinum and proposed it as an intermediate in the synthesis of cellulose. They withdrew that proposal when they were unable to demonstrate that extracellular enzymes in cultures of Acetobacter xylinum could convert this polymer into cellulose.

They did not consider the possibility that removal of the sidegroups might occur at the membrane during secretion of the polymer. In addition, Delmer has discussed observations by a number of investigators wherein radiolabeled soluble polymers associated with cellulose eventually appeared to have the radiolabel incorporated into the cellulose. Delmer notes that it was not possible to develop convincing evidence that the soluble fractions were precursors to cellulose. In our speculation, we envision the de-branching of the precursor polymer as occurring at the sites of assembly complexes that have been visualized in association with the deposition of cellulose. The particulate complexes are viewed as sites where side branches are removed and cellulose molecules are aligned and organized for deposition as fibrils with the lattice structure of cellulose I.

The microtubules of the cytoskeleton could then be regarded as part of the system for transport of the cellulose precursor polymer to the membrane, and for coordinating movement of the assembly complexes at the surface to regulate the direction of deposition of cellulose fibrils.

The question remains as to where the synthesis of the proposed cellulose precursor can occur. The most likely location is in Golgi systems within the cell. These would be topologically more simple than the Golgi producing the heteropolysaccharides, and would have to be coupled, in some way, with the microtubules.

Distributed Synthesis

From a systems theoretic perspective, the proposals concerning the three major constituents of the cell wall matrix presented above have two common elements. The first is transformation of the solvation characteristics of precursors or intermediates in the synthesis. The second is introduction of multiple points of regulation of the processes of biogenesis. Transformation of hydrophobic structures into soluble ones by glycoslylation is well recognized in other biological systems.

In the present study, it is most obvious for the monolignols, which occur as the glycosides in the intracellular environment and are deglycosylated prior to polymerization in the course of lignification. It is not as obvious in the case of polysaccharides, but the â-1,4 linked homopolymers of all of the common pyranoses are essentially insoluble in water beyond the heptamers or octamers.

It is also well known that limited amounts of neutral substitution, can dramatically increase solubility of the homopolymers in aqueous media; the xyloglucans are an example of this effect. With cellulose and many of the secondary wall hemicelluloses, it is obvious that their function requires them to be insoluble, for that is the form in which they occur. It is quite plausible that the same mechanism as the one used for the solubilization and subsequent modification of hydrophobic species would also be adapted for the insoluble structural polysaccharides.

With respect to susceptibility to regulation, the availability of multiple points of control is important. To the extent that processes of distributed synthesis are based on activity of multiple enzyme systems at different points in the cycle of synthesis, the opportunities for regulation of assembly through expression of encoded processes are expanded.

This is important because in the formation of tertiary structure many degrees of freedom at the molecular level of each of the constituents must be controlled. That is, many aspects of structure at the nanoscale level must be coordinated. This is obvious from examination of walls in tangentially adjacent cells in a particular tissue; what they have in common goes beyond the primary structures of the constituents of the walls.

The common features include, among others, the lateral dimensions and orientation of cellulose fibrils in their walls, the patterns of inter-unit linkages in the lignin and the sequence of deposition of the hemicelluloses in the layers of the secondary wall.

The regulation of these aspects of structure argues for activity of agents of expression of encoded characteristics. The entry into distributed synthesis of multiple enzyme systems, separated with respect to activity both in time sequence and in location, is the most plausible mechanism for expression of this encoding.

In presenting this perspective we do not minimize the importance of inherent self organizing characteristics of polysaccharides in the development of tertiary structure. Rather we note that such self organizing tendencies must be modulated by a higher level of coordinating information if self organization of the constituents in adjacent cells is to proceed in the coherent manner reflected by similarities in the molecular architecture of adjacent cells in particular tissues.

HIERARCHIC ORGANIZATION IN SECONDARY THICKENING

The hierarchic nature of structure in plants is reflected in anatomical descriptions and in studies of differentiation and morphogenesis at the cellular level. The hierarchy of structures has not been examined, however, from an information-theoretic perspective with respect to the constraints that one level of organization provides with respect to structures and processes at lower levels.

In the present work our concern is with the constraints that the symmetry of an annual ring implies with respect to processes of assembly of the individual cells. The cross-section of an annual ring in a tree trunk, in the absence of any reaction wood, possesses a cylindrical symmetry that points to synchrony of the processes of assembly within all cells at the circumference. This, in turn, suggests a coupling of the processes within the different cells.

The symmetry is expressed at two levels of organization. The more obvious one is reflected in the geometry of cells as established during morphogenesis, and is an integral part of discussions of organization in plant anatomy. The next level of symmetry, and the one that is central in the present study, is that governing the synchrony of processes of deposition of cell walls after their final geometry has

been established during earlier phases of morphogenesis. This level of symmetry is not precisely geometric; it is defined by similarity in the molecular architecture of cell walls when specified in terms of patterns of deposition of cellulose, the composition of hemicelluloses, and patterns of interunit linkages in lignin. It has been suggested that morphogenetic processes proceed as though each cell has a map and an internal clock to regulate its development in relation to its function and location. In annual rings it appears that the clocks of cells formed at the same point in time are kept in synchrony by some external mechanism.

The result is symmetry along the circumference. This in turn suggests that radial growth proceeds with very similar regulation of the cell wall deposition along the perimeter in the cambial zone. The consequences are readily visualized with the aid of Figure, which shows an SEM image of the corner of a sample of loblolly pine sectioned horizontally, radially and tangentially.

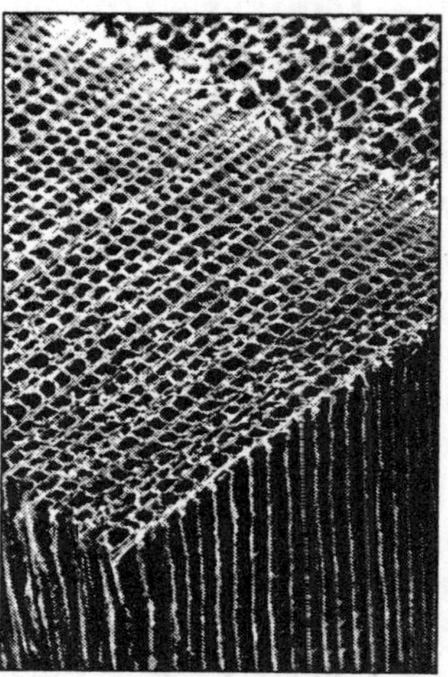

Fig. An SEM Image of a Corner of Loblolly Pine Tissue Sectioned Horizontally

Figure shows that though there may be minor differences between cells deposited along the perimeter, these subtle differences are preserved within the series of daughter cells from the same initial. Thus it is clear that the formation of cell walls must be regulated through the response of intracellular processes that reflect the expression of different parameters for the cell wall in response to signal molecules released through the action of environment-sensing systems in foliage and meristems.

Moreover, these processes that appear to be synchronized must be the primary determinants of molecular organization in the cell wall. That this is so is reflected by the fact that no difference has ever been reported in the molecular organization of wood on the basis of northern or southern solar exposure; clearly the side of the tree with the highest solar exposure would be expected to have, on average, a higher temperature during daylight hours.

If microenvironment and its influence on kinetics were primary determinants of structure, such differences should have been observed decades if not centuries ago. The conclusion then must be that the organization of the cell wall is regulated by intracellular functions that are synchronized. In studies of secondary thickening in woody species reviewed by Steeves and Sussex, it was shown that mechanical pressure has a significant influence on cell wall organization. However, pressure cannot act alone in regulating morphogenesis because its modulation cannot possibly carry the level of information that is necessary to orchestrate the many processes involved in the patterning of molecular architecture.

There are many aspects of structure that could be examined with respect to synchrony of their development within an annual ring. In relation to secondary walls, three come immediately to mind; the fibril angles within different layers, the radial distribution of hemicelluloses across the layers, and the organization of lignin within the polysaccharide matrix.

The variation of fibril angles within different layers of secondary walls and the relative uniformity of this variation within an annual ring are well recognized. So also is some gradual variation of fibril angles from one annual ring to the next as the tree matures. The implications for cellulose synthesis and deposition are broad and well beyond the scope of the present report. It can be said, however, that any mechanism proposed for the formation of cellulose must allow for synchrony of expression of genetically encoded aspects of structure, even as it also allows for variation of this expression at different stages of development. A high degree of orchestration of intracellular, membrane and extracellular processes within adjacent cells is also implied.

The distribution of hemicelluloses across secondary walls in woody tissue has been investigated for a number of species. It is implicit in all of these studies that the distributions are uniform within the same tissue types and within annual rings. In relation to the hemicelluloses, the question of uniformity has not arisen because their basic identity is established within the Golgi apparatus. Further limited modification in the extracellular environment does not depart from intracellular control, since the population of extracellular glycosidases is the result of intracellularly regulated processes.

The organization of lignin is the one aspect of structure that has heretofore been assumed to be dominated by chemical microenvironment within the polysaccharide matrix. The variables defining microenvironment include the levels of peroxidase and/or laccase activity, the rate of generation of hydrogen peroxide or the availability of oxygen, the local pH levels, and the rates of infusion of precursors into the matrix.

But the dominant parameter in any system governed by the kinetics of chemical reactions is temperature. The symmetry of the annual ring, in spite of significant temperature variations during daylight hours, essentially excludes the possibility that temperature is a determinant of structure.

This, in turn, suggests that, while microenvironment may influence the progress of formation of covalent bonds between phenyl propane units, it cannot control patterns of inter-unit linkages. These are clearly regulated by intracellular processes that are sufficiently coupled that they can be synchronized among the cells undergoing lignification simultaneously.

Regulation and synchrony are most likely to occur through the provision of similar templates for assembly of lignin in cells that are forming their cell walls contemporaneously. A question then arises as to the nature of the templates. The considerations presented in the next section point to hemicelluloses as the most plausible carriers of the information.

HEMICELLULOSES AS TEMPLATES FOR THE ASSEMBLY OF LIGNIN

In biological systems the primary carriers of information are molecular. It is therefore necessary to consider which molecules involved in the process of wood cell wall assembly are the most plausible carriers of structural information. Cellulose is the key skeletal component in cell walls and, although its pattern of association varies among species and among different tissues within the same species it is a constant in cell wall tissue. The hemicelluloses, on the other hand, vary from one tissue type to another, and, within an individual cell wall, can vary with radial position.

Furthermore, they are modified whenever the tissue is subjected to external stress, and this modification is correlated with differences in the structure of lignin formed under the same conditions. Finally, their assembly within the Golgi apparatus is governed by precisely encoded information. Thus, they are capable of providing the information linkage between intracellular and extracellular processes.

The next question is whether they have the capacity to induce organization in the structure of lignin. Much information has been accumulated to support an affirmative answer to this question. The most significant is the effect of polysaccharide matrices on synthetic lignin analogs. Molecular modeling studies also have shown that the monomers and oligomers of lignin can associate with polysaccharide surfaces and confirm findings of Komamine et al. and Haigler et al. who observed an association of lignification with cellulose and xylan in differentiating tracheary elements from cultures of Zinnia elegans.

The remaining question concerns the manner in which information is carried and transferred. Since oligosaccharides are known to be involved in biological recognition processes, and since the side branches of most hemicelluloses tend to be mono-, di- or tri-saccharides, it is plausible to regard them as the key. When one also considers that oligomers of lignin, formed by membrane bound enzymes, occur in the extracellular environment, it does seem likely that specific oligosaccharide branches are programmed to bind specific monomers or oligomers of lignin and to organize them in space in preparation for their polymerization by radical coupling or propagation reactions.

The β-1,4 linked backbones of cellulose and the hemicelluloses may provide sites for binding of the monomers of lignin into ordered arrays, the evidence for which was outlined above.

We would note here that our observations of the influence of hemicelluloses on the structure of cellulose formed in their presence indicate that hemicelluloses also play a role in regulation of the aggregation of cellulose. They appear to be key factors in governing the distribution between Iβ and Iâ forms. Thus, the role of the hemicelluloses appears to be definition of the overall character of the polysaccharide matrix.

This is another instance where extracellular aggregation is under regulation of intracellular processes through control of the identity of the regulatory molecules. It is viewed as an intermediate point in the formation of the cell wall, prior to onset of lignification, but in preparation for it. The influence on the structure of lignin is then regarded as the second stage in the regulatory function of hemicelluloses. Further consideration of these questions is well beyond the scope of the present report.

In conclusion, we believe that analysis of the molecular organization of the constituents of plant cell walls in their native state is an important key to advancing understanding of biogenesis of the plant cell wall matrix. We have examined issues associated with aggregation of the constituents and suggested plausible pathways for the assembly of the matrix that are based on distributed synthesis of the major polymeric constituents.

We then considered molecular organization as the first stationary level of expression of phenotypic form and examined the implications of the similarity of this expression within contemporaneous cells in an annual ring of secondary growth. We concluded that all aspects of molecular organization must be governed by intracellular processes that can be orchestrated at levels well beyond those of the individual cells.

We have proposed that the primary molecular carriers of organizing information, that is the mediators of the orchestration, are the hemicelluloses, and we have suggested the manner of their action in the formation of the structure of lignin. The results of our analysis of hierarchic organization have a number of additional implications with respect to structure and its formation that are beyond the scope of the present report.

One that is worthy of note at this time is that the organization of the deposition of the secondary wall, and its systematic variation within different tissues in higher plants, requires intimate coupling and orchestration of intracellular, membrane and extracellular processes, at levels higher than that of the individual cell in a manner that, to our knowledge, has not heretofore been considered.

It is also our view that after consolidation of the cell wall, the hemicelluloses continue to play a very important role within the walls. That is to regulate and maintain hydration of the cell wall constituents and act as a counterbalance to the many hydrophobic secondary metabolites that are formed as wood matures. However, this function of the hemicelluloses is beyond the scope of this report.

Chapter 7

Future of Stem Cells

Tissue engineering is an interdisciplinary field that applies the principles and methods of bioengineering, material science, and life sciences toward the assembly of biologic substitutes that will restore, maintain, and improve tissue functions following damage either by disease or traumatic processes. The general principles of tissue engineering involve combining living cells with a natural/synthetic support or scaffold to build a three-dimensional living construct that is functionally, structurally and mechanically equal to or better than the tissue that is to be replaced. The development of such a construct requires a careful selection of four key materials:

1) scaffold,
2) growth factors,
3) extra cellular matrix,
4) cells.

Much progress has been made in the field of tissue engineering, but further work toward organ and tissue replacement is necessary. The optimal cell source, scaffold design, and in vitro bioreactors, the use and development of micro fabrication technology to create vascularized tissues and organs are still being investigated. The search for and use of an appropriate multipotent or pluripotent stem cell in tissue engineering is an emerging concept. Certainly, many areas of stem cell research and their potential clinical applications are associated with controversies; therefore, it is important to address the ethical, legal, and social issues early.

This paper will provide an overview of tissue engineering and stem cells, and describe the current progress with stem cell research in tissue engineering, the potential implications on medical treatment and the economic impact with the passage of Proposition.

Artificial transplantation or transplanted organs is a successful therapy for otherwise incurable end-stage diseases or tissue loss. However, such interventions are challenged by organ shortage, the necessity of lifelong immunosuppression and its potential for serious complications. Tissue engineering has emerged as a rapidly expanding approach to address these problems and is a major component of regenerative medicine. Tissue engineering is an interdisciplinary field that applies the principles and methods of bioengineering, material science, and life sciences toward the assembly of biologic substitutes that will restore, maintain, and improve tissue functions following damage either by disease or traumatic processes.

The general principles of tissue engineering involve combining living cells with a natural/synthetic support or scaffold that is also biodegradable to build a threedimensional living construct that is functionally, structurally and mechanically equal to or better than the tissue that is to be replaced. The development of such a construct requires a careful selection of four key components:

- Scaffold,
- Growth factors,
- Extracellular matrix (ECM),
- Cells.

Scaffold materials are three-dimensional tissue structures that guide the organization, growth and differentiation of cells. Scaffolds must be biocompatible and designed to meet both nutritional and biological needs for the specific cell population. Growth factors are soluble peptides capable of binding cellular receptors and producing either a permissive or preventive cellular response toward differentiation and/or proliferation of tissue. ECM must be capable of providing the optimal conditions for cell adhesion, growth, and differentiation within the construct by creating a system capable of controlling environmental factors such as pH, temperature, oxygen tension, and mechanical forces.

These conditions are determined by the particular cell lines and the properties of the scaffold. Finally, the development of a viable construct involves a suitable supply of cells that are ideally nonimmunogenic, highly proliferative, easy to harvest, and have the ability to differentiate into a variety of cell types with specialized functions. There are two primary methods to harvest cells and culture. In cases where direct harvest is not feasible, as seen in many patients with extensive end-stage organ failure or cells with limited proliferative capacity in culture, stem cells are envisioned as being an alternative source of cells.

Despite the growing interest in embryonic stem cell research, progress has been hampered by ethical and legislative debates. The political, ethical, and religious opposition toward embryonic stem cell research, which primarily uses discarded nontransferred human embryos for their derivation, have biased research toward adult stem cells and severely restricted federal funding in the United States.

In the mist of such controversies, California has become the leading state to support stem cell research by passing Proposition or the California Stem Cell Research and Cures Initiative. It promises to provide California a possible solution to bridge critical research funding gaps by committing $3 billion for stem cell research and to potentially accelerate the understanding of their therapeutic potential. It does not however support human reproductive cloning. The purpose of this paper is to provide an overview of tissue engineering and stem cells, and describe the current progress with stem cell research in tissue engineering and the potential implications on medical.

FUNDAMENTALS OF TISSUE ENGINEERING

In 1933, the concept of tissue engineering was first introduced when mouse tumor cells demonstrated survival when encased in a biocompatible polymer membrane and implanted into the abdominal cavity of chick embryos. Several decades later, Chick et al., demonstrated that pancreatic beta cells from neonatal rats, cultured on synthetic capillaries and perfused with medium, released insulin in response to changes in glucose concentration.

In the early 1980s, Burke et al., successfully created artificial skin with fibroblasts seeded onto collagen scaffolds for the treatment of extensive burn injury. Clinically, this is still being utilized today. Efforts are now being undertaken for engineering a variety of tissue and organ types with an emphasis on the application of stem cells. Ultimately, the goal of tissue engineering is to regenerate tissues and restore organ function through cell implantation and matrix incorporation into the patient.

Current approaches to tissue engineering can be stratified into substitutive, histioconductive, and histioinductive. Substitutive approaches (ex vivo) are essentially whole organ replacement, whereas histioconductive approaches (ex vivo) involve the replacement of missing or damaged parts of an organ tissue with ex-vivo constructs. In contrast, histioinductive approaches facilitate self-repair and may involve gene therapy using DNA delivery via plasmid vectors or growth factors.

A number of criteria must be satisfied in order to achieve effective, long-lasting repair of damaged tissues.

- An adequate number of cells must be produced to fill the defect.
- Cells must be able to differentiate into desired phenotypes.
- Cells must adopt appropriate three-dimensional structural support/scaffold and produce ECM.
- Produced cells must be structurally and mechanically compliant with the native cell.
- Cells must successfully be able to integrate with native cells and overcome the risk of immunological rejection.
- There should be minimal associated biological risks.

The source of cells utilized in tissue engineering can be autologous (from the patient), allogenic (from a human donor but not immunologically identical), or xenogenic (from a different species donor). Autologous cells represent an excellent source for use in tissue engineering because of the low association with immune complications. Autologous cells are however not cost effective and batch controlled for universal clinical use. In contrast, allogenic cells offer advantages over autologous cells in terms of uniformity, standardization of procedure, quality control and cost effectiveness.

Cell sources can be further delineated into mature (non-stem) cells, adult stem cells or somatic stem cells, embryonic stem cells (ESCs), and totipotent stem cells or zygotes. The utility of mature cells is restricted because of its low proliferative and differentiating potential. Adult stem cells are resident stem cells found in specific niches or tissue compartments and are important in maintaining the integrity of tissues like skin, bone and blood.

They are undifferentiated cells that can be programmed to differentiate into specific tissue types. At least 20 major categories of somatic stem cells have been identified in mammals. They can be found in bone marrow, blood, cornea and retina of the eye, dental pulp, liver, skin, GI tract, and pancreas. ESCs are derived from the inner cell mass of the pre-implantation blastocyst. They are undifferentiated, immature cells that are capable of unlimited self-renewal with the ability to differentiate.

Stem cells (3 types)

Rapid advances are being made in stem cell research with a focus on their therapeutic potential for regenerative medicine and other biomedical applications. In combination with tissue engineering, stem cells hold a number of promises in further advancing contemporary medicine. Traditionally, adult stem cells were believed to form a small number of cells restricted to a particular germ layer origin; however, some evidence now indicates that adult stem cells isolated from various tissues have greater plasticity than previously thought. Several researchers have attributed this apparent plasticity of adult stem cells to developmental signals-mediated differentiation. For example, Azizi et al., demonstrated that marrow stromal stem cells transplanted into the rat brain can acquire a neural phenotype.

Neural and muscle-derived stem cells isolated from cloned mice were able to differentiate into all cell types of hematopoietic lineage. A year later Clark et al, reported that murine neural stem cells could differentiate into all three germ layers with appropriate culture modifications. In 2000 Lagasse et al., demonstrated that the injection of murine bone marrow stem cells into a mutant mouse line with progressive liver failure were able to form hepatocytes and restore liver function.

Recently, however, others reported cell fusion, rather than signal-mediated differentiation, as a possible mechanism for transdifferentiation. For example, Alvarez- Dolado et al., demonstrated the evidence for cell fusion of bone marrow-derived cells with neurons and cardiomyocytes and the lack of transdifferentiation without cell fusion in these tissues, suggesting that cell fusion may be necessary for the observed plasticity of adult stem cells. This finding is not without controversies. Later, others demonstrated that stem-cell plasticity is a true characteristic of neural stem cells, suggesting that transdifferentiation can be accomplished without cell fusion. In summary, the concept of adult stem cell plasticity is new, and the phenomenon is not thoroughly understood.

The ESC is defined by its origin, that of a blastocyst before uterine implantation. Three unique characteristics define primate ESCs:

- They are derived from the four to five day-old embryos that have been produced in an in vitro fertilization clinic.
- They possess a prolonged undifferentiated quality that is capable of self-renewal.
- They demonstrate stable developmental potential to form the derivatives of all three embryonic germ layers ('pluripotent') even after prolonged culture.

Pluripotent stem cells have been derived in mice and primates from pre-implantation embryos and bone marrow stroma. They can be induced to differentiate into all cell types and are able to colonize tissues of interest after transplantation. Human ESCs are mostly obtained from discarded embryos generated in fertility clinics.

In the presence of leukemia inhibitory factor (LIF) or embryonic fibroblast feeder layers, ESCs can be maintained and expanded in an undifferentiated, pluripotent state in vitro almost indefinitely. Upon withdrawal of LIF, ESCs spontaneously differentiate to form distinct embroid bodies, which contain differentiating cells of ectodermal, endodermal and mesodermal lineage. Much of the development in manipulating ESCs have been based on mouse ESC models, and there are numerous examples of improved tissue function with this usage.

Human ESCs, like no other ESCs, differentiate into somatic cell types that make up the human body. The potential benefits to health care are enormous, ranging from generating neurons for treating patients to Parkinson's disease to learning about the molecular processes that drive tumor development.

Current progress

The current utilization of stem cells in tissue includes skin, blood vessels, cartilage, heart tissue, liver, pancreas and neural tissue. Skin defects are primarily treated by the use of epidermal and dermal constructs. Dermal fibroblasts are obtained from neonatal foreskin, expanded in vitro, seeded onto a scaffold of polylactic or polyglycolic acid before being cultured in a bioreactor system to generate a dermal layer. A bilaminate construct is produced by coating the dermal layer with multiple layers of keratinocytes. The long-term viability of these skin grafts depends on the population of the engrafted skin stem cells. Various techniques have been utilized to promote neovascularization. More recently, cell-based therapies known as endothelial progenitor cells or angioblasts have been utilized.

The lack of vascularity renders cartilage less susceptible to ischemic insult than tissues with greater dependency on direct vascularity. Autologous cell products have been used effectively for cartilage repair. Alternatively, both adult stem cells and ESCs are being investigated for their potential to differentiate into the appropriate cells for repair of damaged bone, cartilage, and tendon.

The complexity and specialized conducting infrastructure of the heart have posed a challenge to duplicate in engineering systems. Furthermore, such effort is further limited by a low proliferative potential of cardiomyocytes. Available engineered heart products are biocompatible, non-biodegradable mechanical. These devices are ineffective for longterm replacement.

Nonetheless, the possibility of development of an engineered heart is exemplified by the successful manufacturing of tissue-engineered valves and myocardial infarct scar remodeling. Combining the present engineering efforts with the potential clinical applications of stem cells may accelerate the rate of the development of these engineered products. Kehat et al., first described the generation of a reproducible spontaneous cardiomyocyte differentiating system from human ESCs. Injection of bone marrow cells in the contracting cardiac wall bordering myocardial infarcts has been demonstrated to regenerate myocardium. The development of human ESC lines and their ability to differentiate to cardiomyocytes holds great promise.

Nevertheless, bridging the gap between the theoretical potential of this unique system and any actual clinical applications will require much work. Presently, liver transplantation is the only successful treatment modality for end-stage liver failure. The limited availability of donor livers has turned our attention to alternative sources including tissue engineering.

A successful construct employs the use of highly porous biodegradable discs to deliver hepatocytes. This approach is limited by the low proliferative potential of hepatocytes and the suboptimal cell survival from the poor access to vascularity for nutrient provisions and removal of metabolic waste. Several types of stem cells, such as human adult hematopoietic and mesenchymal stem cells, have been demonstrated to differentiate into hepatocytes under the appropriate environment, and hence may serve as a unlimited cell source for therap.

Diabetes may be potentially treated by generating insulin-producing cells from stem cells and grafting them into the pancreas in diabetic patients. A much better understanding is necessary for the mechanisms that govern the expansion and differentiation into beta cells. Clinical islet cell transplantation for the treatment of type 1 diabetes has been introduced with enthusiasm.

The procedure is considered minimally invasive and potentially offers the possibility of being performed under donor-specific tolerant conditions. Recently, multipotent precursor cells from the adult mouse pancreas have been identified, and may be a promising candidate for cellbased therapeutic strategies.

Unlike isolated beta cells which release insulin in a monophasic, all-or-none manner without any fine-tuning for intermediate concentrations of blood glucose, pancreatic precursor cells may be cultured to produce all the cells of the islet cluster in order to generate a population of cells that will be able to coordinate the release of the appropriate amount of insulin to the physiologically relevant concentrations of glucose in the blood.

The potential of ESCs in the treatment of brain diseases is also under active investigation. Ma et al., recently demonstrated the first CNS stem cell-derived functional synapse and neuronal network formation on a three-dimensional collagen gel.

Mesenchymal stem cells are ideal for replacing the dead neural cells due to their strong proliferative capacity, easy acquisition, and considerable tolerance of genetic modifications. Following the transduction of brain-derived neurotrophic factor (BDNF) via recombinant retroviral vectors into the human mesenchymal stem cells, nearly all cells were demonstrated to express BDNF, and the quantity of BDNF in the culture medium was increased by approximately 20,000-fold.

Another area under investigation is genetic manipulation

of adult stem cells as a potential therapeutic strategy to surmount the problem of organ shortage. Oncology studies have identified numerous genes involved in controlling cell growth. By introducing growth-promoting genes, many stem cells can be immortalized and gain the ability to differentiate. For example, immortalized human hepatocyte stem cells grafted into rats with acute liver failure can keep the animals alive. There is however some concern regarding oncogenic risks with using growth-promoting genes. This problem may be unlikely by introducing three levels of safeguards:

- Deletion of the transferred oncogene by site-specific recombination followed by different selection,
- Allogenic transplantation requiring temporary immunosuppression,
- Introduction of a "cellsuicide" gene that can be activated by administering a drug.

While the selection of appropriate cell source remains a key to the successful creation of stable, complex three-dimensional constructs, equally important is the design of robust scaffolds and ECM that mimics a native physiologic environment and a capillary network in a three-dimensional structure. In collaboration with Draper Laboratories, Shieh et al., have developed a computational model of vascular circulation.

Limitations

One of the major challenges of the field is to design an appropriate capillary network to allow gas exchange, provide nutrition, and remove metabolic waste from the implanted constructs. It will be necessary to create a tubular system with a minimum diameter of 10 ìm to allow vascular ingrowth, especially important for multilayered solid organ constructs with a volume greater than 2 to 3 cubic mm. An alternative is to implant a vascularized construct in an ectopic, well-vascularized location, such as omentum.

Special attention will be given to the control of the local milieu of each cell type as different cell types often require different culturing environments, making it difficult to design a multilayered organ system construct. A challenge will be to create a scaffold capable of supporting various cell types. The appropriate expression sequence, dosing, and duration for growth factors also needs to be defined if we are to maximize ESCs utility in tissue growth and maintenance in the pluripotent state.

Ethical/technical considerations

There are three main research programmes in stem cell research.

- Research on adult stem cells.
- Research on ESCs from discarded embryos that are produced via in vitro fertilization.
- Research on ESCs obtained via therapeutic cloning.

There is little controversy surrounding human adult stem cells. On the other hand, human ESCs face great controversies, the extent of which is dependent on the type of research programmes.

Various autologous adult stem cells, including mesenchymal, hematopoietic, neural, muscle, and hepatic stem cells, are actively being investigated. They are more likely immunocompatible and are not associated with ethical concerns unlike their ESC counterparts. Challenges remain in regard to optimization of isolation techniques that minimize contamination, permanently maintain the desired cell types following differentiation, and increase the production of a large number of cells that are adequate for the construction of a tissue or organ.

Furthermore, much of the data cited have not been reproducible. The utility of adult stem cells would effectively avoid the ethical problem of using ESCs. But much needs to be learned about adult stem cells in regard to their replication and differentiation. ESCs may offer technical advantages over other sources of somatic precursors. For example, ESCs can be generated in abundant quantities in the laboratory and can be grown in their undifferentiated state for many generations.

In contrast, researchers have had difficulty finding

laboratory conditions under which some adult stem cells, especially hematopoetic stem cells from blood or bone marrow, can proliferate without becoming specialized. This technical barrier to proliferation has limited the ability of researchers to explore the capacity of certain types of adult stem cells to generate sufficient numbers of specialized cells for transplantation purposes.

Unlike adult stem cells, ESCs can be grown indefinitely in tissue culture. There is an increased risk for immune rejection because of genetic mismatch, which in turn would demand lifelong high-dose immunosuppressant treatment. Furthermore, the creation of human ESCs derived from discarded, non-transferred human embryos using an immunosurgical technique has resumed arguments that have not been resolved.

A SART-RAND study identified at least 400,000 frozen embryos in storage since the late 1970s. Only 2.8 per cent of these embryos have been designed for research. With a 15 per cent efficiency rate for generating a human ESC line from blastocysts and incorporating some losses to the freeze-thaw process, only approximately 275 human ESC lines can be created from the excess. The pluripotential virtue of stem cells itself renders limitations: the efficiency of differentiation appears to be compromised for certain cell types, such as hepatocytes and cardiomyocytes, and the quality of differentiation is still questionable.

The use of ESCs as an important cell source for tissue engineering has great potential for the treatment of diseases, but there are alternatives, such as the use of adult stem cells and ESCs from therapeutic cloning under active investigation. Worldwide legislation reflects the dilemma, ranging from the more permissive to the nominally more restrictive (the United States—only minimal federal financial funds granted to research that meets the strict NIH guidelines, but stem cell research is legal without any restriction in the private sector) to the truly restrictive (Germany, Austria, Ireland, France—laws on reproductive medicine ban the extraction from stem cells from a human embryo).

The host immune response to allogenic embryonic stem cells poses a major challenge in addition to ethical and religious concerns regarding ESCs. This has led researchers to investigate creation of a universal donor cell by making histocompatible proteins on the cell surface thus reducing the cells' antigenicity.

This is the heart of nuclear transfer technology or "therapeutic cloning". However, the technology has its own shortcomings. It has a low efficiency rate which in turn requires a large number of oocytes. Its inability to control differentiation requires a better understanding of ESC growth, differentiation, and application in engineering tissues. Furthermore, therapeutic cloning protocols in vitro culture systems must be optimized before contemplating the use of this technology for cell therapy.

All three research programmes discussed here are directed at increasing our knowledge about basic cell biology, creating new therapies through stem cell culture and control of cell differentiation, and producing commercially viable stem cell products either by the direct patenting of stem cell lines or by combining stem cell technologies with genetic engineering or other patentable interventions.

The primary discussion on stem cells is concerned with the ethical issues raised by each of these programmes and with whether these issues should influence regulatory decisions regarding public financial support for research.

Proposition passage in California

The California Stem Cell Research and Cures Initiative, recently passed by statewide ballot in November 2004. As a result, the California Institute for Regenerative Medicine (CIRM or Institute) was created in order to provide breakthrough cures for devastating diseases. These diseases include, but are not limited to, diabetes, heart diseases, cancer, osteoporosis, multiple sclerosis, Alzheimer's disease, Parkinson's disease, Lou Gerig's disease and spinal cord injuries by utilizing tax-free state bonds to fund stem cell research.

The institute would ensure that strict ethics codes will be adopted and enforced, including utilizing guidelines developed by the National Institutes of Health. Mirroring the existing state law, the institute would strictly prohibit human reproductive cloning.

In addition to finding cures for various diseases, the goal of the institute is also to reduce the rapidly rising healthcare costs in California and boost its economy. Proposition is designed not to burden taxpayers, authorizing tax-free state bonds that will provide an average of $295 million per year over ten years to support stem cell research in California. It assumes that a large share of Health care costs is caused by diseases and injuries that can potentially be cured with stem cell therapies.

Opponents to Proposition have argued that the bill places an unnecessary and unwise financial burden on an already-insufficient state budget. According to some estimates, state bond debt, which amounted to $33 billion in May of 2004, might rise to more than $50 billion by July of 2005, largely due to the costs of Proposition.

In addition, the initiative has yet to assure the Californian taxpayers whether the state will actually receive any royalties from discoveries paid for by state dollars. Questions about significant conflicts of interest on the Independent Citizen's Oversight Committee, (the "board of directors" of CIRM) and even about the scientific integrity of the stem-cell lines available for research have been raised.

In spite of this opposition, many defend the economic merits of Proposition. As no payments would be demanded within the first five years of the bill's implementation, the state would remain free from financial stress and allowed time to balance its budget. No tax increases would accompany the legislation.

The potential effects that the passage of California's Proposition would have on other states and the nation as a whole are not to be underestimated. New Jersey, the nation's first state to create a publicly funded institute for stem cell research and the second state to pass legislation legalizing the research, has found it necessary to review its policy and is considering an increase in their research grants.

Other states, including Massachusetts, Maryland, Wisconsin and Connecticut, are also jumping in with proposals to fund research cloning.

These changes are happening against the recent passage of the "declaration" of banning all forms of human cloning by the United Nations. A recent Swiss law that mirrors laws in most European countries, allows medical research on stem cells taken from human embryos but bars cloning.

Britain and Belgium allow embryo creation for research. Britain has an independent embryo authority to scrutinize and monitor all projects and has issued only two cloning licenses since 2001.

What has made the United States such a fertile ground for expanding embryo research is not its liberal laws but the lack of them. Congress has tried but failed to pass legislation largely because of irreconcilable differences over when life begins.

Both Presidents Bill Clinton and George W. Bush banned the National Institutes of Health on ethical grounds from funding the creation of human research embryos, although their orders apply only to federally funded work. In 2001, Bush additionally narrowed NIH research to embryonic stem cells already harvested from spare IVF embryos. These restrictions in turn triggered an influx of private and state funding, and the dramatic Proposition.

Future of tissue engineering with stem cell research

Although much progress has been made in the field of tissue engineering, further work toward organ and tissue replacement is necessary. The optimal cell source, scaffold design, and in vitro bioreactors, the use and development of microfabrication technology to create vascularized tissues and organs are still being investigated.

The search for and use of an appropriate multipotent or pluripotent stem cell in tissue engineering is an emerging concept.

Certainly, many areas of stem cell research and their potential clinical applications are associated with controversies; therefore, it is important to address the ethical, legal, and social issues early. Many technical questions are yet to be answered and require close interdisciplinary collaborations of surgeons, engineers, chemists, and biologists, with the ultimate goal of functional tissue restoration.

As more scientific knowledge will be gained from stem cell research, hopefully, some of the current ethical and technical concerns will be answered or removed in the future.

Chapter 8

Disease Transmission

An infectious disease is a clinically evident disease resulting from the presence of pathogenic microbial agents, including pathogenic viruses, pathogenic bacteria, fungi, protozoa, multicellular parasites and aberrant proteins known as prions. These pathogens are able to cause disease in animals and/or plants.

Infectious pathologies are usually qualified as contagious diseases (also called communicable diseases) due to their potentiality of transmission from one person or species to another. Transmission of an infectious disease may occur through one or more of diverse pathways including physical contact with infected individuals. These infecting agents may also be transmitted through liquids, food, body fluids, contaminated objects, airborne inhalation, or through vector-borne spread.

The term infectivity describes the ability of an organism to enter, survive and multiply in the host, while the infectiousness of a disease indicates the comparative ease with which the disease is transmitted to other hosts. An infection however, is not synonymous with an infectious disease, as an infection may not cause important clinical symptoms or impair host function.

Classification

Among the almost infinite varieties of microorganisms, relatively few cause disease in otherwise healthy individuals. Infectious disease results from the interplay between those few pathogens and the defenses of the hosts they infect. The appearance and severity of disease resulting from any pathogen depends upon the ability of that pathogen to damage the host as well as the ability of the host to resist the pathogen. Infectious microorganisms, or microbes, are therefore classified as either primary pathogens or as opportunistic pathogens according to the status of host defenses.

Primary pathogens cause disease as a result of their presence or activity within the normal, healthy host and their intrinsic virulence (the severity of the disease they cause) is, in part, a necessary consequence of their need to reproduce and spread. Many of the most common primary pathogens of humans only infect humans, however many serious diseases are caused by organisms acquired from the environment or which infect non-human hosts.

Organisms which cause an infectious disease in a host with depressed resistance are classified as opportunistic pathogens. Opportunistic disease may be caused by microbes that are ordinarily in contact with the host, such as pathogenic bacteria or fungi in the gastrointestinal or the upper respiratory tract and they may also result from (otherwise innocuous) microbes acquired from other hosts (as in Clostridium difficile colitis) or from the environment as a result of traumatic introduction (as in surgical wound infections or compound fractures).

An opportunistic disease requires impairment of host defenses, which may occur as a result of genetic defects (such as Chronic granulomatous disease), exposure to antimicrobial drugs or immunosuppressive chemicals (as might occur following poisoning or cancer chemotherapy), exposure to ionizing radiation, or as a result of an infectious disease with immunosuppressive activity (such as with measles, malaria or HIV disease). Primary pathogens may also cause more severe disease in a host with depressed resistance than would normally occur in an immunosufficient host.

One way of proving that a given disease is "infectious", is to satisfy Koch's postulates, which demands that the infectious agent be identified only in patients and not in healthy controls and that patients who contract the agent also develop the disease. These postulates were first used in the discovery that Mycobacteria species cause tuberculosis. Koch's postulates cannot be met ethically for many human diseases because they require experimental infection of a healthy individual with a pathogen produced as a pure culture. Often, even diseases that are quite clearly infectious do not meet the infectious criteria.

For example, Treponema pallidum, the causative spirochete of syphilis, cannot be cultured in vitro - however the organism can be cultured in rabbit testes. It is less clear that a pure culture comes from an animal source serving as host than it is when derived from microbes derived from plate culture. Epidemiology is another important tool used to study disease in a population. For infectious diseases it helps to determine if a disease outbreak is sporadic (occasional occurrence), endemic (regular cases often occurring in a region), epidemic (an unusually high number of cases in a region), or pandemic (a global epidemic).

TRANSMISSION

An infectious disease is transmitted from some source. Defining the means of transmission plays an important part in understanding the biology of an infectious agent and in addressing the disease it causes. Transmission may occur through several different mechanisms. Respiratory diseases and meningitis are commonly acquired by contact with aerosolized droplets, spread by sneezing, coughing, talking, kissing or even singing.

Gastrointestinal diseases are often acquired by ingesting contaminated food and water. Sexually transmitted diseases are acquired through contact with bodily fluids, generally as a result of sexual activity. Some infectious agents may be spread as a result of contact with a contaminated, inanimate object (known as a fomite), such as a coin passed from one person to another, while other diseases penetrate the skin directly.

Transmission of infectious diseases may also involve a "Vector". Vectors may be mechanical or biological. A mechanical vector picks up an infectious agent on the outside of its body and transmits it in a passive manner. An example of a mechanical vector is a housefly, which lands on cow dung, contaminating its appendages with bacteria from the feces and then lands on food prior to consumption. The pathogen never enters the body of the fly.

In contrast, biological vectors harbor pathogens within their bodies and deliver pathogens to new hosts in an active manner, usually a bite. Biological vectors are often responsible for serious blood-borne diseases, such as malaria, viral encephalitis, Chagas disease, Lyme disease and African sleeping sickness. Biological vectors are usually, though not exclusively, arthropods, such as mosquitoes, ticks, fleas and lice. Vectors are often required in the life cycle of a pathogen. A common strategy, used to control vector borne infectious diseases, is to interrupt the life cycle of a pathogen, by killing the vector.

The relationship between virulence and transmission is complex and has important consequences for the long term evolution of a pathogen. Since it takes many generations for a microbe and a new host species to co-evolve, an emerging pathogen may hit its earliest victims especially hard.

It is usually in the first wave of a new disease that death rates are highest. If a disease is rapidly fatal, the host may die before the microbe can get passed along to another host. However, this cost may be overwhelmed by the short term benefit of higher infectiousness if transmission is linked to virulence, as it is for instance in the case of cholera (the explosive diarrhea aids the bacterium in finding new hosts) or many respiratory infections (sneezing and coughing create infectious aerosols).

PREVENTING TRANSMISSION

One of the ways to prevent or slow down the transmission of infectious diseases is to recognize the different characteristics of various diseases. Some critical disease characteristics that should be evaluated include virulence, distance traveled by victims and level of contagiousness. The human strains of Ebola virus, for example, incapacitate its victims extremely quickly and kills them soon after. As a result, the victims of this disease do not have the opportunity to travel very far from the initial infection zone.

Also, this virus must spread through skin lesions or permeable membranes such as the eye. Thus, the initial stage of Ebola is not very contagious since its victims experience only internal hemorrhaging. As a result of the above features, the spread of Ebola is very rapid and usually stays within a relatively confined geographical area. In contrast, Human Immunodeficiency Virus (HIV) kills its victims very slowly by attacking their immune system. As a result, a lot of its victims transmit the virus to many others before even realizing that they are carrying the disease. Also, the relatively low virulence allows its victims to travel long distances, increasing the likelihood of an epidemic.

Another effective way to decrease the transmission rate of infectious diseases is to recognize the effects of small-world networks. In epidemics, there are often extensive interactions within hubs or groups of infected individuals and other interactions within discrete hubs of susceptible individuals. Despite the low interaction between discrete hubs, the disease can jump to and spread in a susceptible hub via a single or few interactions with an infected hub.

4Thus, infection rates in small-world networks can be reduced somewhat if interactions between individuals within infected hubs are eliminated. However, infection rates can be drastically reduced if the main focus is on the prevention of transmission jumps between hubs. The use of needle exchange programs in areas with a high density of drug users with HIV is an example of the successful implementation of this treatment method. Another example is the use of ring culling or vaccination of potentially susceptible livestock in adjacent farms to prevent the spread of the foot-and-mouth virus in 2001.

DIAGNOSIS AND THERAPY

Diagnosis of infectious disease sometimes involves identifying an infectious agent either directly or indirectly. In practice most minor infectious diseases such as warts, cutaneous abscesses, respiratory system infections and diarrheal diseases are diagnosed by their clinical presentation. Conclusions about the cause of the disease are based upon the likelihood that a patient came in contact with a particular agent, the presence of a microbe in a community and other epidemiological considerations.

Given sufficient effort, all known infectious agents can be specifically identified. The benefits of identification, however, are often greatly outweighed by the cost, as often there is no specific treatment, the cause is obvious, or the outcome of an infection is benign.

Specific identification of an infectious agent is usually only determined when such identification can aid in the treatment or prevention of the disease, or to advance knowledge of the course of an illness prior to the development of effective therapeutic or preventative measures. For example, in the early 1980s, prior to the appearance of AZT for the treatment of AIDS, the course of the disease was closely followed by monitoring the composition of patient blood samples, even though the outcome would not offer the patient any further treatment options. In part, these studies on the appearance of HIV in specific communities permitted the advancement of hypotheses as to the route of transmission of the virus.

By understanding how the disease was transmitted, resources could be targeted to the communities at greatest risk in campaigns aimed at reducing the number of new infections. The specific serological diagnostic identification and later genotypic or molecular identification, of HIV also enabled the development of hypotheses as to the temporal and geographical origins of the virus, as well as a myriad of other hypothesis. The development of molecular diagnostic tools have enabled physicians and researchers to monitor the efficacy of treatment with anti-retroviral drugs.

Molecular diagnostics are now commonly used to identify HIV in healthy people long before the onset of illness and have been used to demonstrate the existence of people who are genetically resistant to HIV infection. Thus, while there still is no cure for AIDS, there is great therapeutic and predictive benefit to identifying the virus and monitoring the virus levels within the blood of infected individuals, both for the patient and for the community at large.

Methods of Diagnosis

Diagnosis of infectious disease is nearly always initiated by medical history and physical examination. More detailed identification techniques involve the culture of infectious agents isolated from a patient. Culture allows identification of infectious organisms by examining their microscopic features, by detecting the presence of substances produced by pathogens and by directly identifying an organism by its genotype. Other techniques (such as X-rays, CAT scans, PET scans or NMR) are used to produce images of internal abnormalities resulting from the growth of an infectious agent. The images are useful in detection of, for example, a bone abscess or a spongiform encephalopathy produced by a prion.

Microbial Culture

Microbiological culture is a principal tool used to diagnose infectious disease. In a microbial culture, a growth medium is provided for a specific agent. A sample taken from potentially diseased tissue or fluid is then tested for the presence of an infectious agent able to grow within that medium. Most pathogenic bacteria are easily grown on nutrient agar, a form of solid medium that supplies carbohydrates and proteins necessary for growth of a bacterium, along with copious amounts of water.

A single bacterium will grow into a visible mound on the surface of the plate called a colony, which may be separated from other colonies or melded together into a "Lawn". The size, colour, shape and form of a colony is characteristic of the bacterial species, its specific genetic makeup (its strain) and the environment which supports its growth. Other ingredients are often added to the plate to aid in identification.

Plates may contain substances that permit the growth of some bacteria and not others, or that change colour in response to certain bacteria and not others. Bacteriological plates such as these are commonly used in the clinical identification of infectious bacteria. Microbial culture may also be used in the identification of viruses: the medium in this case being cells grown in culture that the virus can infect and then alter or kill. In the case of viral identification, a region of dead cells results from viral growth and is called a "plaque". Eukaryotic parasites may also be grown in culture as a means of identifying a particular agent.

In the absence of suitable plate culture techniques, some microbes require culture within live animals. Bacteria such as Mycobacterium leprae and T. pallidum can be grown in animals, although serological and microscopic techniques make the use of live animals unnecessary. Viruses are also usually identified using alternatives to growth in culture or animals. Some viruses may be grown in embryonated eggs.

Another useful identification method is Xenodiagnosis, or the use of a vector to support the growth of an infectious agent. Chaga's disease is the most significant example, because it is difficult to directly demonstrate the presence of the causative agent, Trypanosoma cruzi in a patient, which therefore makes it difficult to definitively make a diagnosis. In this case, xenodiagnosis involves the use of the vector of the Chaga's agent T. cruzi, an uninfected triatomine bug (subfamily Triatominae), which takes a blood meal from a person suspected of having been infected. The bug is later inspected for growth of T. cruzi within its gut.

MICROSCOPY

Another principle tool in the diagnosis of infectious disease is microscopy. Virtually all of the culture techniques at some point, on microscopic examination for definitive identification of the infectious agent. Microscopy may be carried out with simple instruments, such as the compound light microscope, or with instruments as complex as an electron microscope.

Samples obtained from patients may be viewed directly under the light microscope and can often rapidly lead to identification. Microscopy is often also used in conjunction with biochemical staining techniques and can be made exquisitely specific when used in combination with antibody based techniques.

For example, the use of antibodies made artificially fluorescent (fluorescently labeled antibodies) can be directed to bind to and identify a specific antigens present on a pathogen.

A fluorescence microscope is then used to detect fluorescently labeled antibodies bound to internalized antigens within clinical samples or cultured cells. This technique is especially useful in the diagnosis of viral diseases, where the light microscope is incapable of identifying a virus directly.

Other microscopic procedures may also aid in identifying infectious agents. Almost all cells readily stain with a number of basic dyes due to the electrostatic attraction between negatively charged cellular molecules and the positive charge on the dye.

A cell is normally transparent under a microscope and using a stain increases the contrast of a cell with its background. Staining a cell with a dye such as Giemsa stain or crystal violet allows a microscopist to describe its size, shape, internal and external components and its associations with other cells.

The response of bacteria to different staining procedures is used in the taxonomic classification of microbes as well. Two methods, the Gram stain and the acid-fast stain, are the standard approaches used to classify bacteria and to diagnosis of disease.

The Gram stain identifies the bacterial groups Firmicutes and Actinobacteria, both of which contain many significant human pathogens. The acid-fast staining procedure identifies the Actinobacterial genera Mycobacterium and Nocardia.

BIOCHEMICAL TESTS

Biochemical tests used in the identification of infectious agents include the detection of metabolic or enzymatic products characteristic of a particular infectious agent. Since bacteria ferment carbohydrates in patterns characteristic of their genus and species, the detection of fermentation products is commonly used in bacterial identification. Acids, alcohols and gases are usually detected in these tests when bacteria are grown in selective liquid or solid media.

The isolation of enzymes from infected tissue can also provide the basis of a biochemical diagnosis of an infectious disease. For example, humans can make neither RNA replicases nor reverse transcriptase and the presence of these enzymes are characteristic of specific types of viral infections. The ability of the viral protein hemagglutinin to bind red blood cells together into a detectable matrix may also be characterized as a biochemical test for viral infection, although strictly speaking hemagglutinin is not an enzyme and has no metabolic function.

Serological methods are highly sensitive, specific and often extremely rapid tests used to identify microorganisms. These tests are based upon the ability of an antibody to bind specifically to an antigen.

The antigen, usually a protein or carbohydrate made by an infectious agent, is bound by the antibody. This binding then sets off a chain of events that can be visibly obvious in various ways, dependent upon the test. For example, "Strep throat" is often diagnosed within minutes and is based on the appearance of antigens made by the causative agent, S. pyogenes, that is retrieved from a patients throat with a cotton swab.

Serological tests, if available, are usually the preferred route of identification, however the tests are costly to develop and the reagents used in the test often require refrigeration. Some serological methods are extremely costly, although when commonly used, such as with the "strep test", they can be inexpensive.

Molecular Diagnostics

Technologies based upon the polymerase chain reaction (PCR) method will become nearly ubiquitous gold standards of diagnostics of the near future, for several reasons. First, the catalog of infectious agents has grown to the point that virtually all of the significant infectious agents of the human population have been identified.

Second, an infectious agent must grow within the human body to cause disease; essentially it must amplify its own nucleic acids in order to cause a disease. This amplification of nucleic acid in infected tissue offers an opportunity to detect the infectious agent by using PCR. Third, the essential tools for directing PCR, primers, are derived from the genomes of infectious agents and with time those genomes will be known, if they are not already.

Thus, the technological ability to detect any infectious agent rapidly and specifically are currently available. The only remaining blockades to the use of PCR as a standard tool of diagnosis are in its cost and application, neither of which is insurmountable.

The diagnosis of a few diseases will not benefit from the development of PCR methods, such as some of the clostridial diseases (tetanus and botulism). These diseases are fundamentally biological poisonings by relatively small numbers of infectious bacteria that produce extremely potent neurotoxins.

A significant proliferation of the infectious agent does not occur, this limits the ability of PCR to detect the presence of any bacteria.

Clearance and Immunity

Infection with most pathogens does not result in death of the host and the offending organism is ultimately cleared after the symptoms of the disease have waned.

This process requires immune mechanisms to kill or inactivate the inoculum of the pathogen. Specific acquired immunity against infectious diseases may be mediated by antibodies and/or T lymphocytes. Immunity mediated by these two factors may be manifested by:

- A direct effect upon a pathogen, such as antibody-initiated complement-dependent bacteriolysis, opsonoization, phagocytosis and killing, as occurs for some bacteria,

- Neutralization of viruses so that these organisms cannot enter cells,
- Or by T lymphocytes which will kill a cell parasitized by a microorganism.

The immune response to a microorganism often causes symptoms such as a high fever and inflammation and has the potential to be more devastating than direct damage caused by a microbe.

Resistance to infection (immunity) may be acquired following a disease, by asymptomatic carriage of the pathogen, by harboring an organism with a similar structure (crossreacting), or by vaccination. Knowledge of the protective antigens and specific acquired host immune factors is more complete for primary pathogens than for opportunistic pathogens.

Immune resistance to an infectious disease requires a critical level of either antigen-specific antibodies and/or T cells when the host encounters the pathogen. Some individuals develop natural serum antibodies to the surface polysaccharides of some agents although they have had little or no contact with the agent, these natural antibodies confer specific protection to adults and are passively transmitted to newborns.

MORTALITY FROM INFECTIOUS DISEASES

The World Health Organization collects information on global deaths by International Classification of Disease (ICD) code categories. The top infectious disease killers which caused more than 100,000 deaths in 2002 (estimated). 1993 data is included for comparison.

The top three single agent/disease killers are HIV/AIDS, TB and malaria. While the number of deaths due to nearly every disease have decreased, deaths due to HIV/AIDS have increased fourfold.

Childhood diseases include pertussis, poliomyelitis, diphtheria, measles and tetanus. Children also make up a large percentage of lower respiratory and diarrheal deaths.

Worldwide Mortality due to Infectious Diseases

Rank	Cause of death	Deaths 2002	Percent age of all deaths	Deaths 1993	1993 Rank
N/A	All infectious diseases	14.7 million	25.9%	16.4 million	32.2%
1	Lower respiratory infections	3.9 million	6.9%	4.1 million	1
2	HIV/AIDS	2.8 million	4.9%	0.7 million	7
3	Diarrheal diseases	1.8 million	3.2%	3.0 million	2
4	Tuberculosis (TB)	1.6 million	2.7%	2.7 million	3
5	Malaria	1.3 million	2.2%	2.0 million	4
6	Measles	0.6 million	1.1%	1.1 million	5
7	Pertussis	0.29 million	0.5%	0.36 million	7
8	Tetanus	0.21 million	0.4%	0.15 million	12
9	Meningitis	0.17 million	0.3%	0.25 million	8
10	Syphilis	0.16 million	0.3%	0.19 million	11
11	Hepatitis B	0.10 million	0.2%	0.93 million	6
12-17	Tropical diseases(6)	0.13 million	0.2%	0.53 million	9, 10, 16-18

Historic Pandemics

A pandemic (or global epidemic) is a disease that affects people over an extensive geographical area.

- Plague of Justinian, from 541 to 750, killed between 50 and 60 per cent of Europe's population.

- The Black Death of 1347 to 1352 killed 25 million in Europe over 5 years (estimated to be between 25 and 50% of the populations of Europe, Asia and Africa - the world population at the time was 500 million).
- The introduction of smallpox, measles and typhus to the areas of Central and South America by European explorers during the 15th and 16th centuries caused pandemics among the native inhabitants. Between 1518 and 1568 disease pandemics are said to have caused the population of Mexico to fall from 20 million to 3 million.
- The first European influenza epidemic occurred between 1556 and 1560, with an estimated mortality rate of 20%.
- Smallpox killed an estimated 60 million Europeans in the 18th century alone. Up to 30% of those infected, including 80% of the children under 5 years of age, died from the disease and one third of the survivors went blind.
- The Influenza Pandemic of 1918 (or the Spanish Flu) killed 25-50 million people (about 2% of world population of 1.7 billion). Today Influenza kills about 250,000 to 500,000 worldwide each year.

Emerging Diseases and Pandemics

In most cases, microorganisms live in harmony with their hosts. Such is the case for many tropical viruses and the insects, monkeys, or other animals in which they have lived and reproduced. Because the microbes and their hosts have co-evolved, the hosts gradually become resistant to the microorganisms. When a microbe jumps from a long-time animal host to a human being, it may cease to be a harmless parasite and become pathogenic.

With most new infectious diseases, some human action is involved, changing the environment so that an existing microbe can take up residence in a new niche. When that happens, a pathogen that had been confined to a remote habitat appears in a new or wider region, or a microbe that had infected only animals suddenly begins to cause human disease.

Several human activities have led to the emergence and spread of new diseases:

- Encroachment on wildlife habitats. The construction of new villages and housing developments in rural areas brings people into contact with animals—and the microbes they harbor.
- Changes in agriculture. The introduction of new crops attracts new crop pests and the microbes they carry to farming communities, exposing people to unfamiliar diseases.
- The destruction of rain forests. As countries make use of their rain forests, by building roads through forests and clearing areas for settlement or commercial ventures, people encounter insects and other animals harboring previously unknown microorganisms.
- Uncontrolled urbanization. The rapid growth of cities in many developing countries tends to concentrate large numbers of people into crowded areas with poor sanitation. These conditions foster transmission of contagious diseases.
- Modern transport. Ships and other cargo carriers often harbor unintended "passengers", that can spread diseases to faraway destinations. While with international jet-airplane travel, people infected with a disease can carry it to distant lands, or home to their families, before their first symptoms appear.

MODES OF DISEASE TRANSMISSION

Four routes for the spread of microorganisms exist: contact, airborne, common vehicle and vector-borne. Contact transmission involves direct contact in which body-to-body contact takes place, or indirect in which the susceptible person comes into contact with a contaminated intermediate host (fomite).

Large droplet transmission is judged a form of contact transmission in which large droplets contaminated with microorganisms are generated when an infected person sneezes, coughs, or talks.

These droplets are propelled short distances and deposited on a susceptible host's conjunctiva or mucosa. Airborne transmission happens by aerolisation of an infectious agent through droplet nuclei.

These residual droplets become aerosolised and disperse widely, dependent on environmental conditions and remain suspended in air for indefinite periods. Common vehicle transmission involves one inanimate vehicle, which transmits infection to many hosts and typically applies to microorganisms spread by food and water.

Vector-borne transmission results from the spread of disease by insects and vermin. The infectious diseases that have been transmitted on commercial airlines.

All types of disease transmission are relevant to commercial air travel. Large droplet and airborne mechanisms probably represent the greatest risk for passengers within the aircraft because of the high density and close proximity of passengers.

In addition to proximity, successful spread of contagion to other hosts is dependent on many factors, including infectiousness of the source; pathogenicity of the micro-organism; duration of exposure; environmental con-ditions (ventilation, humidity, temperature) and host-specific factors such as general health and immune status. How these factors affect risk of disease transmission within the aircraft cabin is unclear.

Risk of Transmission

The risk of disease transmission within the confined space of the aircraft cabin is difficult to determine. Insufficient data prohibits meta-analysis, which would allow an idea of the probability of disease transmission for each respective contagion.

Many of the available epidemiological studies are compromised by reporting bias caused by incomplete passenger manifests, thereby complicating risk assessment.

Despite these limitations, data suggest that risk of disease transmission to other symptom-free passengers within the aircraft cabin is associated with sitting within two rows of a contagious passenger for a flight time of more than 8 h.

This association is mainly derived from investigations of inflight transmission of tuberculosis but is believed to be relevant to other airborne infectious diseases. Some variation in this association has been reported, with one outbreak of severe acute respiratory syndrome (SARS) in which passengers seated as far as seven rows from the source passenger were affected.

Risk of disease transmission within the aircraft cabin also seems to be affected by cabin ventilation. In general, proper ventilation within any confined space reduces the concentration of airborne organisms in a logarithmic fashion and one air exchange removes 63% of airborne organisms suspended in that particular space.

The main laminar flow pattern within the aircraft cabin with the practice of frequent cabin air exchanges and use of HEPA filtration for recirculated air clearly limits transmission of contagion.

Transmission becomes widespread within all sections of the passenger cabin when the ventilation system is nonoperational, as shown by an influenza outbreak when passengers were kept aboard a grounded aircraft with an inoperative ventilation system.

Risk assessment incorporating epidemiological data into mathematical models may show how proximity and ventilation affects disease transmission aboard commercial airlines.

Deterministic modelling with data from an in-flight tuberculosis investigation revealed that doubling ventilation rate within the cabin reduced infection risk by half.

Risk also reduced exponentially to almost zero in passengers seated 15 seats from the infectious source. Clearly ventilation provides a crucial determinant of risk and efforts to increase ventilation will reduce risk.

AIRBORNE AND LARGE DROPLET-TRANSMITTED DISEASES

Tuberculosis

Tuberculosis is a serious global threat and estimates suggest that about a third of the world's population has the disease. The transmission of Mycobacterium tuberculosis is the most studied model of the spread of airborne pathogens aboard aircraft. Several studies about in-flight transmission of tuberculosis have been reported, with most being done in the mid-1990s.

Two of the seven investigations revealed a probable link of onboard transmission. In the first occurrence, a flight attendant was the index case and two documented tuberculin skin test conversions occurred during 5 months in 1992 in 212 fellow crew members and 59 frequent flyer passengers. The second and largest, incident was of a passenger with pulmonary tuberculosis travelling from Baltimore to Chicago and then on to Honolulu.

Four of 15 fellow passengers seated within two rows of the index passenger had positive tuberculin skin test conversion. Although there is a risk of tuberculosis within the aircraft cabin, no cases of active disease have been reported as a result of air travel. Transmission within the aircraft cabin seems to be more likely with close proximity to a contagious passenger (within two rows) over a long time (greater than 8 h) and not as a result of the practice of recirculating 50% of the cabin air. An overall probability of infection in the order of one in 1000 when a symptomatic source is present has been suggested and this probability of risk is similar to, if not less than, those in other confined spaces.

SARS

SARS is a non-typical pneumonia caused by a coronavirus. The global spread by air travellers and in-flight spread of SARS has been documented. The disease is believed to usually be spread by large aerosolised droplets or by direct and indirect contact, but airborne or small droplet transmission better explains the distribution of SARS cases that has occurred on commercial airlines. Evidence suggests that transmission of SARS during the Amoy Gardens outbreak in Hong Kong was a result of airborne spread via a viral plume.

A total of 40 flights have been investigated for carrying SARS-infected passengers. ive of these flights have been associated with probable on-board transmission of SARS in 37 passengers. Most of those passengers were seated within five rows of the index case. One 3-hour flight carrying 120 passengers travelling from Hong Kong to Beijing on March 15, 2003, began a superspreading event accounting for 22 of the 37 people who contracted SARS after air travel.

Laboratory-confirmed SARS coronavirus infection occurred in 16 people, two passengers had a diagnosis of probable SARS and four were reported to have SARS but could not be interviewed.

The number of secondary cases from that flight remains under investigation, but more than 300 people might have been affected. This pattern could be important because it did not follow the typical example of in-flight transmission of airborne pathogens—ie, risk of disease transmission is associated with a flight time of more than 8 h and sitting within two rows of the index passenger.

The duration of the Hong Kong to Beijing flight was 3 h and affected passengers were seated seven rows in front and five rows behind the index passenger. Possible explanations for this outbreak distribution include airborne transmission rather than direct contact spread; a malfunctioning cabin filtration system; and passengers infected before or after the flight. No on-board transmissions have occurred since late March, 2003, when the WHO issued specific guidelines for in-flight containment of SARS.

As the first severe contagious disease of the 21st century, SARS exemplifies the everpresent threat of new infectious diseases and the real potential for rapid spread made possible by the volume and speed of air travel. Finally, the distribution pattern of SARS transmission aboard the flight emphasises the need to study airborne transmission patterns aboard commercial aircraft.

Common Cold

Common cold outbreaks as a result of air travel have not been reported, which could be attributable to the difficulties of investigating such outbreaks in view of the ubiquitous nature of the common cold.

One study compared the risk of developing an upper respiratory tract infection during air travel in passengers flying on aircraft that recirculated 50% cabin air versus aircraft using 100% fresh air in the passenger cabin.36 Recirculation of aircraft cabin air was not a risk factor for contracting upper respiratory tract infection symptoms.

Influenza

The aircraft as a vector for global spread of influenza strains is a greater concern than is in-flight transmission. The fact that influenza outbreaks worldwide have been affected by influenza strains imported by air travel is well established; however, only three studies of in-flight transmission of influenza have been reported. The first was in an outbreak of influenza A/Texas strain aboard a commercial carrier in 1979 that resulted in 72% of all passengers aboard the airline contracting influenza within 72 h. The secondary attack rate in their families was estimated to be 20% within 2 weeks. The high transmission rate in this particular case was believed attributable to passengers being kept aboard the aircraft for 3 h with an inoperative ventilation system while repair work was being done.

The second study described an outbreak of influenza A/Taiwan/1/86 at a naval air station in 1989 in military personnel who were returning from temporary duty.35 Transmission of influenza occurred both on the ground and aboard two DC-9 aircraft that transported the squadron from Puerto Rico to a Florida naval station.

The third outbreak happened in 1999 in mine workers travelling on a 75-seat aircraft. 15 passengers travelling with the index case developed symptoms within 4 days. Nine of the 15 were seated within two rows and all were seated within five rows, of the index case. No further influenza outbreaks aboard commercial aircraft have been reported since 1999.

Other Airborne Diseases

Meningococcal disease occurs after direct contact with respiratory secretions and is associated with high morbidity and mortality rates. A case of meningococcal disease associated with air travel is defined as the development of the illness within 14 days of travel on a flight lasting at least 8 h, including ground delay, take off and landing. The US Centers for Disease Control received 21 reports of suspected air travel-associated meningococcal disease from 1999 to 2001.

In all cases the index person was contagious while aboard a commercial flight, but no secondary cases of the disease were reported. People seated next to an ill passenger should be quickly contacted and given chemoprophylaxis within 24 h of identifying the index case. Chemoprophylaxis given more than 14 days after onset of illness in the index case is probably of little or no value.

Measles is an airborne and highly contagious viral infection with an attack rate of about 80%. Transmission can occur during the prodromal illness and passengers might be unaware of their diagnosis at the time of travel. Measles is no longer endemic in the USA and importations from developing countries account for most outbreaks. Imported measles and associated cases accounted for at least 17% of all reported cases within the USA during 1982. From 1996 to 2000 30% of all imported measles cases were estimated to be in people who flew while symptomatic with the disease.

Three case studies have described measles transmission during commercial air travel. A case report in 1982 identified seven secondary cases of measles epidemiologically linked to the index case as a result of in-transit exposure; one was a passenger flying on the same aircraft as the index case and five others were people who had visited at least one common departure gate with the index case. In the same year, another study reported an index passenger who infected two fellow passengers on a flight from Venezuela to Miami. A third report recorded eight cases of in-flight transmitted measles during a 10-h flight from New York to Tel Aviv in 1994.

The source case was not identified but was speculated to be a crew member. In 2004, a passenger with measles travelled from Japan to Hawaii, but this did not result in any transmission to other air travellers. International adoptions have a significant role in the number of imported measles cases.

COMMON VEHICLE DISEASES

The most commonly reported diseases transmitted on aircraft have been spread by the fecal-oral route via contaminated food. A total of 41 in-flight food-borne outbreaks resulting in 11 deaths were documented between 1947 and 1999.

Salmonella is the most usually reported food-borne pathogen spread via a commercial airline, with fifteen documented outbreaks between 1947 and 1999 infecting nearly 4000 passengers and resulting in seven deaths. Eight food-borne outbreaks caused by Staphylococcus and one associated death were reported between 1947 and 1999.

One of the largest cases involved 57% of the passengers served a ham omelette on an international flight in 1975. Surprisingly, only one viral-induced enteritis outbreak has been described. In this incident, contaminated orange juice was the vehicle of transmission and a Norwalk-like agent was isolated from fecal samples of 30 ill passengers.

There have been a few reported cases of ill passengers as a result of food or water contaminated with Vibrio cholerae consumed during international air travel. The first documented in-flight outbreak was in 1972 on a flight from London to Sydney via Singapore. Of the 47 people who developed cholera, which was attributed to a cold appetiser served during the flight, one died. The largest outbreak of airline-associated cholera occurred in 1992, during a cholera epidemic in Latin America. During a flight from Buenos Aires via Lima to Los Angeles, 75 passengers developed cholera, resulting in ten passengers being admitted and one death. A cold seafood dish, prepared in one of the cholera-affected countries, was implicated as the source of transmission.

No food-borne or water-borne outbreaks have been reported over the past 5 years, which is probably attributable to greater use of prepackaged frozen meals, improved food handling and inspection, but might represent under-reporting by passengers or reporting bias.

VECTOR-BORNE AND ZOONOTIC DISEASES

Vector-borne diseases are very frequent causes of morbidity and mortality in many parts of the world and the potential of their importation via commercial aircraft remains a risk. Many cases of malaria occurring in and around airports all over the world in people who had not travelled to endemic areas, known as airport malaria, is evidence that malaria-carrying mosquitoes can be imported on aircraft. A total of 87 cases of airport malaria have been reported, 75 of which happened in Europe.

Dengue and yellow fever are both transmitted by mosquitoes of the genus Aedes. The Aedes mosquito has been introduced into countries where it had not previously been present and many of these mosquitoes were likely to have spread by aircraft.

A case of dengue fever was reported in Germany in a couple returning from a trip to Hawaii.71 Airport transmission of dengue fever was suspected in this particular case. The disinsection of aircraft—spraying aircraft before landing to kill insects and vector control around airports, as well as immunisation requirements, seem to have been effective in preventing outbreaks in non-endemic areas.

Although International Health Regulations recommend disinsecting aircraft travelling from countries with malaria and other vector-borne diseases, only five countries do so. Common and exotic animals are regularly transported on aircraft and may carry disease. Many zoonotic pathogens cause emerging and reemerging diseases. Up to now, no zoonotic outbreaks associated with air travel have been reported;

however, continued monitoring of air transport of animals, especially from developing countries, is needed.

Bioterrorism Agents

The potential for spread of bioterrorism agents via air travel exists. A well written, comprehensive review of bioterrorism agents and the implications of air transport is available and a new interest in previously eradicated smallpox has emerged. In-flight spread of this disease has been documented and so the potential for in-flight spread of smallpox is of public health concern. In 1963 an epidemic was reported in Sweden, in which the index case was attributed to in-transit exposure, either at the air terminal or on the aeroplane and caused 24 secondary cases and four deaths.

Viruses causing viral haemorrhagic fever, such as ebola and lassa, have also been the focus of media attention and have been investigated for potential transmission while aboard an aircraft. Although aerosol spread of ebola has not been documented in man, this mode of transmission occurs in non-human primates. Lassa, known to be transmitted via large droplets, is thought to have an incubation period of up to 3 weeks, making infectious passengers potentially symptom-free and unaware of their status at the time of travel.

A study of passengers exposed to an index case of lassa fever in-flight noted no evidence of transmission, even in the 19 passengers seated within two rows of the index passenger. Because both lassa and ebola viruses have frequent fatal outcomes and no vaccine is available, appropriate infection control procedures should be followed to prevent the transmission of these diseases. Lassa fever is treatable with ribavirin if the drug is given within the first 6 days of illness, making early diagnosis especially important.

Management of Infectious Disease

Early recognition and appropriate infection control measures are needed when passengers become exposed to an infectious or potentially infectious passenger. Government and international laws provide legal authority to control the movement of passengers with communicable diseases. This authority ranges from issuing travel alerts to quarantine of passengers arriving at airports.

Although air carriers have the right to refuse to take passengers who are ill with a communicable disease or medically unfit for air travel, systematic screening of passengers for contagious diseases and excluding passengers with infectious symptoms is impractical.

Health care professionals are expected to identify individuals who are unfit for air travel, or advise the flying public of how to safely travel by air. Prevention of a disease outbreak is the most important means of control and travellers should therefore be advised to postpone any air travel when they are ill.

Good hand hygiene has been proven to reduce the risk of disease transmission and air travellers should make it part of their normal travel routine. Although masks play a crucial part in infection control in health care settings, their use is unproven in disease control within the aircraft cabin.

However, a mask should be placed on a passenger suspected of having SARS and the passenger should be isolated. US CDC and WHO guidelines exist, on when and how to notify passengers and flight crew after they have been exposed to infectious diseases aboard commercial aircraft. Briefly, the airline is consulted whenever a health department determines that a passenger is infectious at time of flight. The airline then notifies passengers and flight crews in writing. Notification is typically limited to flights longer than 8 h and, in some cases, dependent upon the design of the aircraft, to passengers seated only in the same cabin area.

Health officials have access to passenger manifests, but these lists are frequently incomplete or unavailable, making it difficult to locate potentially exposed passengers. Contact information in one large investigation into inflight tuberculosis transmission was inaccurate for 15% of passengers. Although air carriers are under no obligation to archive passenger manifests, most have internal policies to do so for up to 3 months.

RESPIRATORY DISEASES AND DISORDERS

Common Respiratory Ommon Respiratory

There is a common misperception that older persons tend to overestimate or exaggerate respiratory symptoms; however, the opposite is more often true. For example, many older persons and their physicians tend to underestimate the importance of dyspnea, which may go undiagnosed until advanced disease is evident. This is partly due to the fact that dyspnea is blamed on deconditioning and age.

Older persons will often adjust their activity level to compensate for insidiously shrinking lung function and disabling dyspnea. Such changes in life style often go unnoticed by family, the patient's physician and even the patient. Pulmonary or cardiac disorders, or both, may underlie such modifications in life style and testing (eg, pulmonary function tests or chest radiography) may reveal major abnormalities such as asthma, emphysema, or pulmonary fibrosis.

Another complicating feature of symptom recognition in older persons is that older persons often have more than one explanation for their problems.

A patient may have overlapping symptoms of dyspnea, cough and wheezing because of a combination of diseases such as asthma or emphysema, obstructive sleep apnea, heart failure and gastroesophageal reflux.

DYSPNEA

Dyspnea becomes prominent in end-stage lung diseases such as chronic obstructive pulmonary disease (COPD) and idiopathic pulmonary fibrosis. Importantly, the level of dyspnea is the best predictor of quality of life, yet it does not correlate with either oxygenation or pulmonary function tests. A thorough history and physical examination can help tailor both testing and empirical treatment choices.

For example, in an older person presenting with dyspnea and associated nocturnal cough, one would first consider common diseases such as asthma, emphysema, allergic rhinitis with postnasal drip and gastroesophageal reflux disease.

Minimal testing (eg, pulmonary function tests only) followed by an empiric trial directed toward the most likely cause would be a reasonable approach.

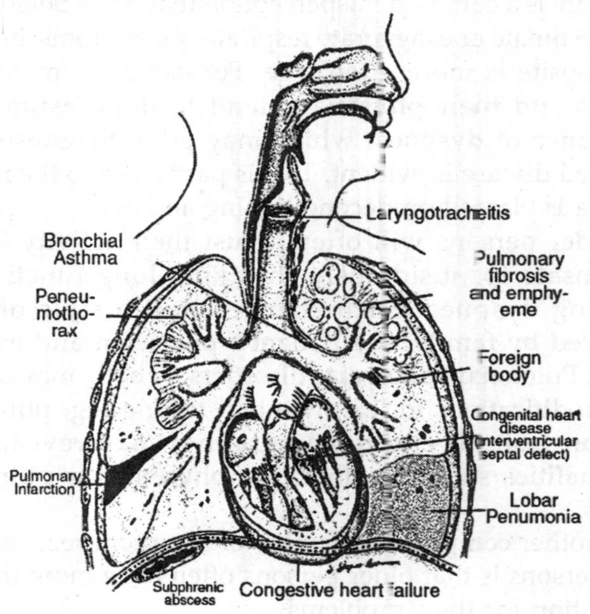

Fig. Dyspnea

In the same patient, the presence of significant weight loss or constitutional symptoms (fever, night sweats) could suggest other diseases, such as malignancy or tuberculosis.

At times, the particular language the patient choses to describe the dyspnea can be revealing, such as "heavy" for cardiac dysfunction or deconditioning or "Tight" for angina or asthma. The common causes of dyspnea in older persons to consider include COPD, cardiac disease, asthma, interstitial lung disease and deconditioning.

CHRONIC COUGH

Fortunately, most patients can be reassured that chronic cough, though particularly annoying, usually has a benign cause. By far, the most common causes of chronic cough are postnasal drip, asthma and gastroesophageal reflux.

These three diagnoses account for over 90% of the causes identified in most series and a reasonable approach to the treatment of chronic cough, then, is empiric treatment for these conditions. Not infrequently, a combination of these conditions may contribute, so treatment for multiple causes may be warranted when single therapies are ineffective.

Less common yet important differential diagnostic considerations of cough in older persons would include drug effects (eg, angiotensin-converting enzyme inhibitors), heart failure, laryngeal dysfunction, Bordetella pertussis infection, chronic cough after viral upper respiratory tract infection or secondary bacterial infections, recurrent aspiration, or respiratory tract abnormalities such as bronchiectasis or airway tumors.

WHEEZING

Though asthma is a common cause of wheezing in all age groups, it is not the most common cause, particularly if the wheezing is not associated with cough or dyspnea. Postnasal drip is another common cause to consider; also, the rates of heart failure rise in the older age groups and associated pulmonary edema may present as "cardiac asthma." Finally, airway hyperresponsiveness from chronic bronchitis is not uncommon in older patients with a history of wheezing and sputum and tobacco use.

PULMONARY DISEASES ASTHMA

After childhood, there is a second peak in the prevalence of asthma beyond the age of 65; 5% to 10% of older persons meet criteria for obstruction and bronchial hyperreactivity. The rate of death from asthma has increased most significantly in those aged 65 and over, accounting for up to 45% of all asthma deaths. This is likely due to reduced awareness of bronchial constriction on the part of the patient (with delays in seeking medical attention), as well as under-recognition and undertreatment on the part of clinicians.

Therapy of asthma in older and younger persons differs in several ways. Paramount to the care of the elderly asthmatic patient is adequate instruction in the proper use of peak expiratory flow monitoring (because of the older person's decreased perception of bronchoconstriction) and in the correct activation of the metered-dose inhaler.

Neurologic, muscular and arthritic diseases in older persons can lead to suboptimal timing and discoordination in the actuation of the inhaler device. The clinician should observe the patient actually using the inhaler. Inhaled corticosteroids (or other controller drugs such as leukotriene receptor antagonists) represent the mainstay of therapy in both older and younger persons.

Use the lowest effective dose and counsel regarding rinsing of the oropharynx to avoid thrush. In older people, theophylline is fraught with adverse effects and drug interactions and should be considered a third-line drug to be used only as a once-daily medication in the evening for severe asthma or COPD, targeting a serum level of 10 to 12 mg/dL if tolerated. Although controversy persists as to whether the response to relievers such as â-agonists varies with age, these drugs remain a mainstay as an as-needed reliever medication.

The potential for adverse effects of â-agonists—for example, hypokalemia or possible QT prolongation in cardiac patients on digoxin or other medications—warrants adequate controller use by asthmatic patients to minimize their overreliance on the â-agonist. The use of long-acting â-agonists is controversial and should be considered with caution and only in those patients who are reliably able to manage medications.

CHRONIC OBSTRUCTIVE PULMONARY DISEASE

COPD affects approximately 15 million people in the United States and is the fourth most common cause of death after heart disease, cancer and stroke. The prevalence and mortality rate from COPD is increasing, especially in older persons. Episodes of acute respiratory failure that require mechanical ventilation are associated with mortality rates ranging from 11% to 46%.

The National Lung Health Education Programme

Executive Committee has noted that the morbidity and mortality from COPD accounts for more than $15 billion per year in U.S. medical care expenditures. Hospitalization continues to represent the largest component of cost for COPD patients. The diagnosis of airflow limitation is challenging in that no single item or combination of items from the history and clinical examination excludes airflow limitation. The finding most strongly associated with a decreased likelihood of airflow limitation is a history of never having smoked cigarettes (especially in patients without a history of wheezing and without wheezing on examination).

Wheezing noted on physical examination is the most potent predictor of airflow limitation and patients with obstructive airflow limitation are 36 times more likely to have wheezing than are patients without this problem. Other findings associated with an increased likelihood of airflow limitation include a barrel-shaped chest, hyperresonance on percussion and a forced expiratory time of greater than 9 seconds measured during the clinical bedside examination.

Smoking cessation at any age has been shown to slow the decline in lung function and aggressive cessation efforts are appropriate even in the oldest-old patient. The basic elements of the approach are the "Five As" from the Agency for Health Care Policy and Research:

- Ask patients about use of tobacco at every office visit.
- Assess readiness to quit.
- Advise patients to quit.
- Assist patients in the quit attempt with aids such as a local cessation programme and pharmacologic agents such as bupropion or nicotine replacement.
- Arrange both a quit date and a follow-up visit

The chief components of daily drug therapy in emphysema consist of a â-agonist, ipratropium bromide or tiotropium, or both drugs in combination. For more severe disease, the use of long-acting â-agonists such as salmeterol along with a combined albuterol and ipratropium bromide metered-dose inhaler can achieve improved adherence and long-term control by reducing the number of inhalers by one (ie, the patient will have only the long-acting inhaler and the combination short-acting inhaler rather than three inhalers).

Inhaled corticosteroids have now been tested in several multicenter randomized controlled trials in COPD with negative results overall; however, there does appears to be a benefit in subgroup analyses of patients with "asthmatic COPD" as defined by spirometry testing with documented bronchodilator responsiveness.

A landmark investigation documented that the use of systemic corticosteroids (intravenous followed by oral) reduces the duration and recurrence of acute exacerbations of COPD for up to 6 months. Importantly, there is no benefit to a course of steroids longer than 14 days. For the few patients (5% to 10%) who do benefit or who require prolonged use of corticosteroids, the risks should be considered and documented in the patient's chart. These risks include peptic ulcer disease, hypertension, cataracts, diabetes mellitus, osteoporosis, psychosis, seizures, poor wound healing, infections and aseptic necrosis of the hip.

Appropriate preventive measures should also be taken in these circumstances of prolonged use, such as using the lowest possible dose of corticosteroids and using supplemental vitamin D, calcium and perhaps a bisphosphonate for those at risk for osteoporosis. Other possible beneficial interventions in older emphysema patients include pulmonary rehabilitation via exercise training and respiratory therapy and education. Both major depression and anxiety have been shown to be present in up to 40% of COPD patients and their treatment must be considered.

In fact, unprovoked anxiety attacks often result in patients' seeking help in emergency departments, being admitted to the hospital and being treated with potentially avoidable courses of oral corticosteroids when the anxiety is not diagnosed and treated.

OBSTRUCTIVE SLEEP APNEA

Sleep-related breathing disorders are very common in older persons and obstructive sleep apnea is the most common type of sleep-related breathing disorder. Obstructive sleep apnea has been associated with cerebrovascular accidents, myocardial infarctions and a threefold increase in mortality. Most patients with obstructive sleep apnea remain undiagnosed and therefore without treatment of this life-threatening, yet potentially correctable disease.

Treatment options include addressing upper-airway obstruction via weight loss, avoidance of alcohol and sedatives, sleeping on one's side or upright, correction of metabolic disorders such as hypothyroidism and continuous positive airway pressure (CPAP) via a nasal mask.

To increase adherence with the use of CPAP, one might order the treatment with "nasal pillows" to increase comfort and "ramping technique" to give a delayed rise in the applied pressure after the patient has fallen asleep. Treatment issues are generally the same for the young and the old and the major consideration for the clinician is a high index of suspicion and clinical recognition of this disease.

IDIOPATHIC PULMONARY FIBROSIS

There are more than 100 causes of restrictive lung diseases; however, the history, examination, serologic testing and biopsy often leave the patient with the diagnosis of idiopathic pulmonary fibrosis. This disease is increasing in prevalence with the aging of our population. Rarely is it an inherited disorder. Pulmonary fibrosis is extremely frustrating for all involved because of its relentless progression. The median survival is 3 to 5 years.

The presentation is normally one of insidious dyspnea (often unrecognized because of a decrease in the activity level on the part of the patient) and cough. Clubbing is often a prominent finding on physical examination in pulmonary fibrosis, as opposed to emphysema, which rarely causes clubbing (prompting a search for another disease such as occult lung cancer).

Oral corticosteroids (0.5 mg/kg/day) for 3 to 6 months is the initial therapeutic maneuver most commonly taken, yet only 10% to 20% of patients respond and adverse effects are often prominent. Early referral to a subspecialist is warranted if the patient wishes to consider further therapeutic attempts so that steroid-sparing agents such as azathioprine or enrollment in a randomized controlled trial of newer pharmacologic agents (eg, interleukin-10 or interferon gamma) can be considered.

PULMONARY THROMBOEMBOLISM

The incidence of pulmonary thromboembolism triples between the ages of 65 and 90 years and has a reported 10% recurrence rate within 1 year. Age above 70 years has been independently associated with missed antemortem diagnosis. Importantly, 10% to 20% of patients with documented pulmonary embolism have an entirely normal blood gas (ie, normal Pao and normal A-a gradient for age).

Age-specific risk factors for pulmonary thromboembolism include hypercoagulability due to increases in fibrinogen, activated protein-C resistance due to factor-V Leiden gene mutation, malignancy, stasis (decreased mobility due to stroke, heart failure, or arthritis), or vessel injury (due to trauma or varicosities). The diagnostic work-up is not different for young and older patients.

Anticoagulants are central to therapy and generally guided by the same principles in younger and older patients. Because of lessened cardiopulmonary reserve, it may be even more important to achieve therapeutic levels of heparinization quickly to avoid major adverse hemodynamic or oxygenation defects in older patients. The trend toward increased utilization of outpatient low-molecular-weight heparin preparations, while achieving anticoagulation with warfarin, is supported by large and appropriately designed randomized controlled trials.

There should be an overlap of approximately 1 to 3 days between heparinization and adequate warfarin therapy with an INR target of 2 to 3. Warfarin interacts with many drugs that are commonly used in the older age group. Studies have inconsistently shown that age itself is a risk factor for bleeding risk with use of warfarin.

Duration of therapy for at least 6 months has been shown to be superior to 3 months and the shorter duration should be used only for those patients with either a specific risk factor that is now removed or for those in whom the risk of prolonged anticoagulant therapy clearly outweighs that of completing 6 months of therapy. Indeed, patients with multiple ongoing risk factors for pulmonary thromboembolic disease are to be considered for anticoagulation therapy for up to 2 years or longer. Recurrent pulmonary thromboembolism is usually treated with lifelong anticoagulation.

INTENSIVE CARE OF THE CRITICALLY ILL

There is widespread belief that many patients do not want aggressive care at the end of life. Moreover, eight of ten health care professionals believe that the treatments they offer to patients are often overly burdensome. More consistency in our attempts to obtain patients' preferences for do-not-resuscitate orders might reduce the extent and burden of aggressive medical care for the elderly age group.

Clinicians receive little training regarding this topic. Data from the Study to Understand Prognoses and Preferences for Outcomes and Risks of Treatments (SUPPORT) showed that physician error rates in approximating patients' preferences for mechanical ventilation increased from 36% for patients younger than 50 up to 79% for patients older than 80. Furthermore, SUPPORT data also showed that age was a predictor of less aggressive care in the intensive care unit (ICU) even after adjusting for severity of illness, gender, race and diagnosis.

A report from a cohort of mechanically ventilated patients in a medical ICU, on the other hand, showed that persons older than 75 years spent the same amount of time on the ventilator as did younger patients but had a lower cost of ICU and in-hospital care.

These outcomes were not explained by differences in mortality, since both groups had similar survival rates. Accordingly, the decision to use mechanical ventilation should not be based on age alone and the appropriate use of ventilatory support in the elderly patient requires further prospective evaluation.

When ICU care and mechanical ventilation are chosen to treat an older person with respiratory failure, special issues related to liberation from the ventilator may be important. In contrast to the general medical ICU population, acute lung injury and acute respiratory distress syndrome patients do appear to have a higher mortality with advancing age. Interestingly, three separate studies of more than 1500 patients have shown that older persons recover from their pulmonary physiologic abnormalities such as hypoxemia and ventilatory disturbances at a rate equal to their younger counterparts after acute lung injury.

These patients do, however, have a particularly difficult time during the latter stages of ventilatory support and require nearly twice as long to be successfully liberated from the ventilator and discharged from the ICU. Often comorbid conditions contribute to these patients' deaths despite the apparent correction of their acute lung injury-induced physiologic disturbances.

Sedative and analgesic medications must be used judiciously and delirium, heart failure, hypothyroidism, electrolyte disturbances, oropharyngeal dysfunction and aspiration should be considered in patients with difficulty liberating from the ventilator. For many older people in good physical condition who succumb to an acute illness, cognitive decline is the main threat to their ability to recover their former functional abilities. For those whose physical activities were already limited, cognitive decline may dramatically worsen after an ICU stay and become the major additional threat to quality of life.

Establishing end-of-life care in critically ill patients is treacherous. Numerous studies have documented the tremendous inaccuracy in formulating the prognosis for patients with serious illnesses. At times, in caring for critically ill patients, it may become apparent to the patient, family and clinician that further intervention would not likely be of substantial benefit, depending on the individual patient's goals, values and hopes. Thus, a strict definition of futility is impossible.

The American Medical Association recommends a standardized "fair process" rather than a strict definition of futility. As much as possible, physicians should base futility decisions on factors such as clinical efficacy of treatment, likelihood of mortality and subsequent quality-of-life considerations rather than on chronologic age alone.

A recent analysis of 6303 ICU-related deaths found that 26% of these deaths occurred in patients receiving full ICU care, including failed cardiopulmonary resuscitation; 24% received full ICU care without cardiopulmonary resuscitation; 14% had life support withheld; and 36% (the single largest group) had life support actively withdrawn. Although there was a wide variation in the practice pattern of different ICUs, the authors concluded that the limitation of life support prior to death is a common practice in teaching ICUs across the country.

In another study of 851 patients receiving mechanical ventilation, 63% were successfully liberated, 17% died while receiving mechanical ventilation and 20% had mechanical ventilation withdrawn. The top four reasons given for withdrawing the ventilator were

- Physicians' perception that the patient did not want life support.
- Physicians' prediction that ICU survival was lower than 10%.
- Physicians' prediction that future cognitive function would be severely impaired, the ongoing need for an inotrope or vasopressors.

The inability for physicians to accurately prognosticate beyond hours of impending death must again be acknowledged as challenging such decisions and should temper concerns about the application of technology in critically ill older patients.

SEXUALLY TRANSMITTED DISEASE (STD)

The Venereal Diseases Act of 1917 defined 3 such diseases.

They were syphilis, gonorrhoea and chancroid. In the UK chancroid is unimportant and often forgotten although it is still troublesome in some parts of Africa. Syphilis ebbed but has been resurgent and the prevalence of gonorrhoea is often taken as an index of the degree of promiscuity in the community.

The generic term for the venereal diseases (VD) was changed to sexually transmitted diseases (STD) and the VD clinics became special clinics as new euphemisms were devised to hide the embarrassment of society. More recently the speciality was called genito-urinary medicine (GUM). They are also referred to as sexually transmitted infections (STI). The range of diseases that would meet that classification is now much wider than the 3 that were named in the early part of the last century.

Epidemiology

After a surge in sexually transmitted diseases during the First World War there was a lull between wars and another surge during the Second World War. When young men are about to go off to war and there is a realistic chance that they may never return, there is a great urge to "take comfort" before embarkation and this may not be under judicious conditions.

The 1960s saw an unprecedented ease in contraception with the ready availability of the pill and a soaring rate of promiscuity, without barrier contraception. The slogan of the young was "make love, not war", judgment was often impaired by chemicals and perhaps there was an unrealistic feeling that antibiotics could cure such diseases. The advent of AIDS was a chilling reminder of the limitation of antibiotics and whilst promiscuity fell, as indicated by the incidence of gonorrhoea, this lull was only temporary.

Modes of Transmission

To make a meaningful analysis of the facts and figures, it is important to look at the various ways by which the diseases may be spread. There may be heterosexual activity.There may be homosexual activity. Not all such diseases are always spread by sexual activity. There may be drug taking activity that leads to transmission of such diseases.

The geographical variation in the importance of various forms of transmission with regard to acquired immune deficiency syndrome. Such means of dissemination are not discrete as people may be bisexual, indulging in sexual activity with both sexes and intravenous drug abusers may turn to prostitution to feed their habit. A number of diseases can be vertically transmitted from mother to child.

Another very important group is those who harbour sexually transmitted diseases but they are unaware that they have the disease or have refused to present for diagnosis and treatment. Such people can account for a very large discrepancy between the number known to have the disease and the number who are actually infected. With regard to HIV infection, they may represent more than 30% of cases. They are a particular problem because, if they are unaware of the infection, they are more liable to spread it.

In many cases, infection remains undiagnosed as the individual is asymptomatic. The disease may be more likely to produce symptoms in one sex than the other, but it remains contagious. A man may readily see a primary chancre of syphilis on his glans whilst a woman is unaware of one on her cervix. Candida may cause vaginal discharge and pruritis vulvae but often in men it causes no symptoms.

SYPHILIS

The origins of syphilis are uncertain, as described in the final section. It has long been known to be a sexually transmitted disease that can also be passed from mother to child. It can also be transmitted by blood transfusion but blood is routinely screened for the disease. Pregnant women are also routinely screened for syphilis.

If not stopped by adequate treatment, it can produce a variety of conditions and over the years the disease became known as the great mimicker. The introduction of penicillin transformed the management of the disease and it declined. In the 1960s around 70% of syphilis was in homosexuals.

There would be no reliable figures about the number who caught in from homosexual practices before that time as such a confession would make the patient liable to imprisonment. In recent decades, syphilis has become resurgent, including in the heterosexual population.

It is not uncommon for it to be acquired at the same time as other sexually transmitted diseases and as it has an incubation period of around 21 days, it may not have presented when other diseases are diagnosed.

It is a disease that is very easy to overlook these days, whereas 50 years ago the great mimicker would have been in almost every differential diagnosis. If a patient has an unexpected reaction to an antibiotic, it is worth bearing in mind that this could be a Jarisch-Herxheimer reaction due to latent syphilis. Another disease that can also give that reaction and can also be difficult to diagnose is Lyme disease.

Gonorrhoea

Gonorrhoea has almost certainly been around for many centuries and possibly back to biblical times although it is impossible to be certain. The incubation period is 3 to 5 days. Men tend to suffer urethral discharge, frequency of micturition and dysuria. Women tend to get purulent vaginal discharge, frequency of micturition and anorectal discomfort.

There may be bartholinitis. Men are usually symptomatic but only around 60% of infected women show the characteristic features. However, they may develop salpingitis and even the Fitz-Hugh-Curtis syndrome. Because of variety of sexual practice, swabs should be taken from the endocervix, the urethra in both men and women, the throat and anus. Anal infection can cause proctitis. A suitable transport medium must be used. Co-infection with Chlamydia is common and should be sought.

A number of complications have been mentioned above, including pelvic inflammatory disease. Arthritis may also occur. It is generally polyarticular and migratory, usually settling in one or two joints. The knees are most often involved, followed by the ankles, wrists, tarsi, MTP joints and dorsal tendon sheaths of the fingers. Infection from the mother's genital tract can cause ophthalmia neonatorum. Treatment is usually a single, high dose of antibiotic. Amoxicillin is no longer favoured and resistance patterns may make ciprofloxacin obsolete in favour of a new generation cephalosporin such as cefixime.

Chancroid

This is the 3rd of the classical "venereal diseases". Chancroid is caused by the Gram-negative bacillus Haemophilus ducreyi. It may be difficult to isolate in culture, but can be seen microscopically as short Gram-negative rods. Chancroid presents as a soft, painful chancre that resembles the lesions of genital herpes. It is endemic in Africa, Asia and South America and is more common in men, particularly uncircumcised men.

HIV may also be present in as many as 60% in Africa. It may facilitate infection with HIV. Other common co-existing infections are syphilis and HSV-2. Painful lymphadenopathy in the groin occurs in 50%. There are a number of treatment regimes including azithromycin 1gram as a single dose and ciprofloxacin 500mg, also as a single dose.

HIV/AIDS

HIV/AIDS has replaced syphilis as the STD to be most feared in that it is both a killer disease and a great mimic. Highly active antiretroviral therapy (HAART) has transformed the prognosis but it remains a disease that has annihilated many people in Africa and, as explained in the article on the subject, unless it is taken very much more seriously in the UK, it will kill many more here too.

Chlamydia

Genitourinary chlamydia infection is a disease whose importance is being recognised much more recently. It can cause PID and infertility in women and non-gonococcal urethritis in men. It is symptomatic in only around 80% or women and 50% of men. There is evidence that a screening programme would be useful and cost-effective. At present there are a series of pilots with the intention to establish a national programme based upon their results by 2007. The recommended treatment is doxycycline 100mg twice daily for 7 days or azithromycin 1g as a single dose.

Non-specific Urethritis

Non-specific urethritis (NSU) is also called non-gonococcal urethritis. It affects males with symptoms of dysuria and possibly urethral discharge. It does not affect females although the implicated organisms affect females in other ways. Usually chlamydia is involved although the term NSU is often reserved for when there is male urethritis but neither the gonococcus nor chlamydia can be isolated. It is usually successfully treated with tetracycline or erythromycin.

Reiter's syndrome is a form of reactive arthritis. It can result from a number of infections, one of which is chlamydia. There is symmetrical oligoarthritis, urethritis and conjunctivitis. It can arise after food poisoning or after a sexually transmitted infection. It is one of the seronegative arthropathies associated with the HLA-B27 antigen.

Lymphogranuloma Venereum

This is a disease caused by a type of chlamydia. It is rare in the UK but the HPA has reported outbreaks amongst homosexual men. It derives its name from painful, swollen inguinal lymph nodes.

HERPES GENITALIS

The herpes simplex virus has two types, HSV1 and HSV2. As a general rule, HSV1 causes cold sores on lips whilst HSV2 causes genital herpes but occasionally their role can be reversed, especially after oro-genital sex. The HSV1 virus is very common in children but the HSV2 is more indicative of sexual transmission.

Genital Warts

They are caused by the human papilloma virus. They

may affect the genitals or anus and are also known as condylomata acuminata. Human papilloma virus and genital.

Carcinoma of Cervix

For many years it has been known that carcinoma of the cervix is associated with an early age of onset of sexual intercourse and multiple partners. It is also more common in smokers. About 95% of cases show evidence of infection with the HPV virus types 16 and 18. Types 31 and 33 may also be implicated. This has led to the proposition that the disease is sexually transmitted.

A vaccine against HPV16 has been developed and it could be in part of the immunisation schedule for all girls at around the age of 12. Needless to say, this has brought outrage from the usual quarters, suggesting that girls will see it as a permit for sexual promiscuity. It has been suggested that it may eventually make cervical smears unnecessary. There is no plan to give it to boys although boys spread the virus to girls. This is reminiscent of the early days when rubella vaccine was given only to girls.

Candidiasis

Candida albicans is a common commensal and may become dominant and cause symptoms when the ecology is upset as after a course of antibiotics. The reason for including it in this list is as a reminder that if a woman who is sexually active gets the infection, it is essential to treat her partner too or they will reinfect each other. It should not normally be seen as one of the STIs. For example, if a child gets a disease that is normally sexually transmitted, there is a very strong inference of sexual abuse but if a girl gets vaginal candida after a course of antibiotics, this is no cause for alarm. Candida is far more likely to cause symptoms in the female than the male.

Viral Hepatitis

Hepatitis A is usually spread by the faeco-oral route whilst hepatitis B and C are most often spread through body fluids, especially amongst intravenous drug abusers. However, the sexual transmission of hepatitis B is well established (as explained in Prevention of Hepatitis B). In England and Wales from 1995 to 2000 laboratory data suggested that the annual incidence of HBV infection was 7.4 per 100,000, with injecting drug use as the commonest cause of transmission.

The number of cases attributed to heterosexual contact was fairly stable, whereas the number of cases in homosexual men decreased. Sexual transmission of the hepatitis C virus is possible but uncommon. Less than 5% of the regular sexual partners of people with HCV infection will become infected. This means that infectivity by the sexual route is fairly low. It does not mean that it does not occur.There are specify guidelines for the management of hepatitis B with HIV and hepatitis C with HIV.

Pelvic Inflammatory Disease

Most cases of pelvic inflammatory disease are acquired from ascending infection from the genital tract. The gonococcus and chlamydia are most commonly involved. There is often mixed infection. It can cause chronic pelvic pain and infertility.

Trichomonas Vaginalis

Trichomonas vaginalis is a flagellated protozoan that may be found in the urethra and genital tract of both men and women. It is asymptomatic in up to 50% of both sexes but may cause vaginal discharge, vulval itching, dysuria, or an offensive odour in women.

Sometimes the presenting complaint is of low abdominal discomfort. In men it may cause urethral discharge and dysuria but discharge is rarely profuse. Vaginal discharge occurs in up to 70% of women, but the classical discharge of frothy yellow occurs in only 10 to 30%.

Vulvitis and vaginitis are reported. Around 10% have no abnormalities on examination. Men may have urethral discharge. Treatment is metronidazole, 400mg twice daily for 5 to 7 days or 2g as a single oral dose.

EPIDIDYMO-ORCHITIS

Epididymo-orchitis is usually sexually transmitted in men of less than 35 years old. Chlamydia trachomatis and Neisseria gonorrhoeae are the common organisms. There is unilateral testicular pain and swelling and sometimes urethral discharge if it was sexually transmitted. A scrotal support and NSAIDs give symptomatic relief but antibiotics should be ceftriaxone 250mg intramuscularly as a single dose or ciprofloxacin 500mg as a single oral dose plus doxycycline 100mg twice daily, orally for 10 to 14 days.

Prevention

Sexually transmitted diseases occur as a result of sexual promiscuity. They do not occur within faithful monogamous relationships or in celibacy. Hence, when such diseases are diagnosed, society's response can be both sanctimonious and punitive. Fortunately, a great deal of pragmatism has prevailed. In the armed forces, contracting a sexually transmitted disease used to be an offence punishable at court marshall. This did not stop servicemen from taking injudicious rest and recreation but it did mean that they would hide the disease if it occurred rather than presenting to the medical officer for treatment.

The social stigma of having a sexually transmitted disease is enormous and the GUM clinics, since their origin, have been discrete and kept anonymity for their patients. Patients are often given a number so that they do not have to present with a name. They tend to have a discrete entrance in a distant part of the hospital and, to ensure compliance, the drugs they dispense are exempt from prescription charges. It is the only part of the NHS that will not routinely inform the patient's GP that he has attended.

Despite this cloak of secrecy and anonymity, they will attempt to trace contacts of those with sexually transmissible diseases. This is a very important role that must require the utmost tact and diplomacy.

Prevention of sexual transmission of diseases must depend upon education and attitude. Getting the content right for sex education is not easy. A number of factors have been identified as contributing to a successful programme. Adverse attitudes to the wearing of condoms must also be overcome. As explained in the article on AIDS, the Roman Catholic Church has much to answer for in terms of spreading disinformation about condoms. They even advocate that if one of a married couple have HIV that it is better for the other partner to catch the disease than for them to use condoms.

HISTORY OF SEXUALLY TRANSMITTED DISEASES

There is uncertainty about the origins of syphilis. Some people claim that it was brought back from America by Columbus' sailors but others argue that it originated in the Old World. An archeological find of the bones of an Essex woman suggested that she had advanced syphilis, between 1300 and 1450. This would predate Columbus' voyage of 1492.

It is often claimed that Henry VIII died of syphilis but the evidence for or against is poor. In those days any ulceration was called pox and syphilis was the great pox. The ulcer on his leg that failed to heal may well have been from type 2 diabetes.

The legendary Giacomo Casanova was said to have invented the condom, fashioned from a sheep's intestine, not as a contraceptive but as protection against syphilis. He may have used them but he was not the first. Condoms appear in some of the pictures of A Harlot's Progress by William Hogarth. This series of paintings was completed in 1731, when Casanova would have been 6 years old. The term syphilis was not part of the English language until 1717, following the translation of an Italian work.

The disease had been the pox and in London rhyming slang, the acquisition of any GUM infection is still referred to as "a dose of the Surrey Docks".

The clap refers to gonorrhoea, from the old French word clapier, meaning a brothel. It was long thought that the two diseases were one. In 1767, John Hunter inoculated a subject with matter of gonorrhoea on the prepuce and glans but the inoculum was from a patient suffering from both diseases and syphilis developed. In 1793, Benjamin Bell, an Edinburgh surgeon confirmed that the diseases were distinct as a result of experimentation on medical students. This was in the days before Local Research Ethics Committees. The eminent Guy's surgeon Sir Asley Cooper recognised the two diseases as distinct and in 1824 he wrote in The Lancet "a man who gives mercury in gonorrhoea deserves to be flogged out of the profession because he must be quite ignorant of the principle in which the disease is cured."

The etymology of gonorrhoea is probably from the Greek for seed flow in that it was assumed that the urethral discharge was semen. Folk etymology suggests an association with the biblical town of Gomorrah. Its equally evil neighbour, Sodom, is the origin for the term for anal intercourse.

In 1864 Parliament passed the Contagious Diseases Act. This legislation allowed policeman to arrest prostitutes in ports and army towns and bring them in to have compulsory checks for venereal disease. If the women were suffering from sexually transmitted diseases they were placed in a locked hospital until cured. It was claimed that this was the best way to protect men from infected women.

Many of the women arrested were not prostitutes but they still were forced to go to the police station to undergo a humiliating medical examination.The traditional treatment for syphilis was heavy metals such as bismuth and mercury. They are of dubious efficacy but highly toxic. Hence the adage "one night with venus and a lifetime with mercury." The discovery that penicillin can treat syphilis has revolutionized its management.

AIDS

Acquired immune deficiency syndrome or acquired immunodeficiency syndrome (AIDS or Aids) is a set of symptoms and infections resulting from the damage to the human immune system caused by the human immunodeficiency virus (HIV). This condition progressively reduces the effectiveness of the immune system and leaves individuals susceptible to opportunistic infections and tumors. HIV is transmitted through direct contact of a mucous membrane or the bloodstream with a bodily fluid containing HIV, such as blood, semen, vaginal fluid, preseminal fluid and breast milk. This transmission can involve anal, vaginal or oral sex, blood transfusion, contaminated hypodermic needles, exchange between mother and baby during pregnancy, childbirth, or breastfeeding, or other exposure to one of the above bodily fluids.

AIDS is now a pandemic. In 2007, an estimated 33.2 million people lived with the disease worldwide and it killed an estimated 2.1 million people, including 330,000 children.

Over three-quarters of these deaths occurred in sub-Saharan Africa, retarding economic growth and destroying human capital. Most researchers believe that HIV originated in sub-Saharan Africa during the twentieth century.

The disease was first identified by the U.S. Centers for Disease Control and Prevention in 1981 and its cause identified by American and French scientists in the late 1980s.

Although treatments for AIDS and HIV can slow the course of the disease, there is currently no vaccine or cure. Antiretroviral treatment reduces both the mortality and the morbidity of HIV infection, but these drugs are expensive and routine access to antiretroviral medication is not available in all countries.

Due to the difficulty in treating HIV infection, preventing infection is a key aim in controlling the AIDS epidemic, with health organizations promoting safe sex and needle-exchange programmes in attempts to slow the spread of the virus.

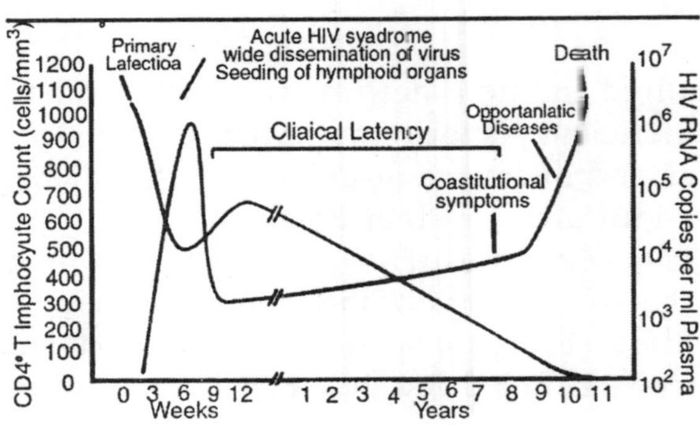

Symptoms

A generalized graph of the relationship between HIV copies (viral load) and CD4 counts over the average course of untreated HIV infection; any particular individual's disease course may vary considerably. CD^{4+} T Lymphocyte count (cells/mm³) HIV RNA copies per mL of plasma

The symptoms of AIDS are primarily the result of conditions that do not normally develop in individuals with healthy immune systems. Most of these conditions are infections caused by bacteria, viruses, fungi and parasites that are normally controlled by the elements of the immune system that HIV damages. Opportunistic infections are common in people with AIDS.

HIV affects nearly every organ system. People with AIDS also have an increased risk of developing various cancers such as Kaposi's sarcoma, cervical cancer and cancers of the immune system known as lymphomas. Additionally, people with AIDS often have spromoting safe sex and needle-exchange programmes in attempts to ystemic symptoms of infection like fevers, sweats (particularly at night), swollen glands, chills, weakness and weight loss. The specific opportunistic infections that AIDS patients develop depend in part on the prevalence of these infections in the geographic area in which the patient lives.

Pulmonary Infections

Pneumocystis pneumonia (originally known as Pneumocystis carinii pneumonia and still abbreviated as PCP, which now stands for Pneumocystis pneumonia) is relatively rare in healthy, immunocompetent people, but common among HIV-infected individuals.

It is caused by Pneumocystis jirovecii. Before the advent of effective diagnosis, treatment and routine prophylaxis in Western countries, it was a common immediate cause of death. In developing countries, it is still one of the first indications of AIDS in untested individuals, although it does not generally occur unless the CD4 count is less than 200 cells per μL of blood. Tuberculosis (TB) is unique among infections associated with HIV because it is transmissible to immunocompetent people via the respiratory route, is easily treatable once identified, may occur in early-stage HIV disease and is preventable with drug therapy.

However, multidrug resistance is a potentially serious problem. Even though its incidence has declined because of the use of directly observed therapy and other improved practices in Western countries, this is not the case in developing countries where HIV is most prevalent. In early-stage HIV infection (CD4 count >300 cells per μL), TB typically presents as a pulmonary disease.

In advanced HIV infection, TB often presents atypically with extrapulmonary (systemic) disease a common feature. Symptoms are usually constitutional and are not localized to one particular site, often affecting bone marrow, bone, urinary and gastrointestinal tracts, liver, regional lymph nodes and the central nervous system.

Gastrointestinal Infections

Esophagitis is an inflammation of the lining of the lower end of the esophagus (gullet or swallowing tube leading to the stomach). In HIV infected individuals, this is normally due to fungal (candidiasis) or viral (herpes simplex-1 or cytomegalovirus) infections. In rare cases, it could be due to mycobacteria. Unexplained chronic diarrhea in HIV infection is due to many possible causes, including common bacterial (Salmonella, Shigella, Listeria or Campylobacter) and parasitic infections; and uncommon opportunistic infections such as cryptosporidiosis, microsporidiosis, Mycobacterium avium complex (MAC) and viruses, astrovirus, adenovirus, rotavirus and cytomegalovirus, (the latter as a course of colitis). In some cases, diarrhea may be a side effect of several drugs used to treat HIV, or it may simply accompany HIV infection, particularly during primary HIV infection.

It may also be a side effect of antibiotics used to treat bacterial causes of diarrhea (common for Clostridium difficile). In the later stages of HIV infection, diarrhea is thought to be a reflection of changes in the way the intestinal tract absorbs nutrients and may be an important component of HIV-related wasting.

Neurological Diseases

Toxoplasmosis is a disease caused by the single-celled parasite called Toxoplasma gondii; it usually infects the brain causing toxoplasma encephalitis but it can infect and cause disease in the eyes and lungs.

Progressive multifocal leukoencephalopathy (PML) is a demyelinating disease in which the gradual destruction of the myelin sheath covering the axons of nerve cells impairs the transmission of nerve impulses.

It is caused by a virus called JC virus which occurs in 70% of the population in latent form, causing disease only when the immune system has been severely weakened, as is the case for AIDS patients.

It progresses rapidly, usually causing death within months of diagnosis. AIDS dementia complex (ADC) is a metabolic encephalopathy induced by HIV infection and fueled by immune activation of HIV infected brain macrophages and microglia which secrete neurotoxins of both host and viral origin. Specific neurological impairments are manifested by cognitive, behavioral and motor abnormalities that occur after years of HIV infection and is associated with low CD^{4+} T cell levels and high plasma viral loads.

Prevalence is 10–20% in Western countries but only 1–2% of HIV infections in India. This difference is possibly due to the HIV subtype in India. Cryptococcal meningitis is an infection of the meninx (the membrane covering the brain and spinal cord) by the fungus Cryptococcus neoformans. It can cause fevers, headache, fatigue, nausea and vomiting. Patients may also develop seizures and confusion; left untreated, it can be lethal.

Tumors and Malignancies

Patients with HIV infection have substantially increased incidence of several malignant cancers. This is primarily due to co-infection with an oncogenic DNA virus, especially

Epstein-Barr virus (EBV), Kaposi's sarcoma-associated herpesvirus (KSHV) and human papillomavirus (HPV).

Kaposi's sarcoma (KS) is the most common tumor in HIV-infected patients. The appearance of this tumor in young homosexual men in 1981 was one of the first signals of the AIDS epidemic.

Caused by a gammaherpes virus called Kaposi's sarcoma-associated herpes virus (KSHV), it often appears as purplish nodules on the skin, but can affect other organs, especially the mouth, gastrointestinal tract and lungs. High-grade B cell lymphomas such as Burkitt's lymphoma, Burkitt's-like lymphoma, diffuse large B-cell lymphoma (DLBCL) and primary central nervous system lymphoma present more often in HIV-infected patients. These particular cancers often foreshadow a poor prognosis. In some cases these lymphomas are AIDS-defining. Epstein-Barr virus (EBV) or KSHV cause many of these lymphomas.

Cervical cancer in HIV-infected women is considered AIDS-defining. It is caused by human papillomavirus (HPV).

In addition to the AIDS-defining tumors listed above, HIV-infected patients are at increased risk of certain other tumors, such as Hodgkin's disease and anal and rectal carcinomas. However, the incidence of many common tumors, such as breast cancer or colon cancer, does not increase in HIV-infected patients.

In areas where HAART is extensively used to treat AIDS, the incidence of many AIDS-related malignancies has decreased, but at the same time malignant cancers overall have become the most common cause of death of HIV-infected patients.

Other Opportunistic Infections

AIDS patients often develop opportunistic infections that present with non-specific symptoms, especially low-grade fevers and weight loss.

These include infection with Mycobacterium avium-intracellulare and cytomegalovirus (CMV) CMV can cause colitis, as described above and CMV retinitis can cause blindness. Penicilliosis due to Penicillium marneffei is now the third most common opportunistic infection (after extrapulmonary tuberculosis and cryptococcosis) in HIV-positive individuals within the endemic area of Southeast Asia.

CAUSE

AIDS is the most severe acceleration of infection with HIV. HIV is a retrovirus that primarily infects vital organs of the human immune system such as $CD4^+$ T cells (a subset of T cells), macrophages and dendritic cells. It directly and indirectly destroys $CD4^+$ T cells.

Once HIV has killed so many $CD4^+$ T cells that there are fewer than 200 of these cells per microliter (μL) of blood, cellular immunity is lost. Acute HIV infection progresses over time to clinical latent HIV infection and then to early symptomatic HIV infection and later to AIDS, which is identified either on the basis of the amount of $CD4^+$ T cells remaining in the blood and/or the presence of certain infections, as noted above. In the absence of antiretroviral therapy, the median time of progression from HIV infection to AIDS is nine to ten years and the median survival time after developing AIDS is only 9.2 months.

However, the rate of clinical disease progression varies widely between individuals, from two weeks up to 20 years. Many factors affect the rate of progression.

These include factors that influence the body's ability to defend against HIV such as the infected person's general immune function. Older people have weaker immune systems and therefore have a greater risk of rapid disease progression than younger people. Poor access to health care and the existence of coexisting infections such as tuberculosis also may predispose people to faster disease progression.

The infected person's genetic inheritance plays an important role and some people are resistant to certain strains of HIV. An example of this is people with the homozygous CCR5-Ä32 variation are resistant to infection with certain strains of HIV. HIV is genetically variable and exists as different strains, which cause different rates of clinical disease progression.

Sexual Transmission

Sexual transmission occurs with the contact between sexual secretions of one person with the rectal, genital or oral mucous membranes of another. Unprotected receptive sexual acts are riskier than unprotected insertive sexual acts and the risk for transmitting HIV through unprotected anal intercourse is greater than the risk from vaginal intercourse or oral sex. However, oral sex is not entirely safe, as HIV can be transmitted through both insertive and receptive oral sex.

The risk of HIV transmission from exposure to saliva is considerably smaller than the risk from exposure to semen, one would have to swallow liters of saliva from a carrier to run a significant risk of becoming infected Sexual assault greatly increases the risk of HIV transmission as protection is rarely employed and physical trauma to the vagina frequently occurs, facilitating the transmission of HIV.

Other sexually transmitted infections (STI) increase the risk of HIV transmission and infection, because they cause the disruption of the normal epithelial barrier by genital ulceration and/or microulceration; and by accumulation of pools of HIV-susceptible or HIV-infected cells (lymphocytes and macrophages) in semen and vaginal secretions. Epidemiologi-cal studies from sub-Saharan Africa, Europe and North America suggest that genital ulcers, such as those caused by syphilis and/or chancroid, increase the risk of becoming infected with HIV by about four-fold.

There is also a significant although lesser increase in risk from STIs such as gonorrhea, Chlamydial infection and trichomoniasis, which all cause local accumulations of lymphocytes and macrophages.

Transmission of HIV depends on the infectiousness of the index case and the susceptibility of the uninfected partner. Infectivity seems to vary during the course of illness and is not constant between individuals. An undetectable plasma viral load does not necessarily indicate a low viral load in the seminal liquid or genital secretions. However, each 10-fold increase in the level of HIV in the blood is associated with an 81% increased rate of HIV transmission. Women are more susceptible to HIV-1 infection due to hormonal changes, vaginal microbial ecology and physiology and a higher prevalence of sexually transmitted diseases. People who have been infected with one strain of HIV can still be infected later on in their lives by other, more virulent strains.

Exposure to Blood-borne Pathogens

This transmission route is particularly relevant to intravenous drug users, hemophiliacs and recipients of blood transfusions and blood products. Sharing and reusing syringes contaminated with HIV-infected blood represents a major risk for infection with HIV.

Needle sharing is the cause of one third of all new HIV-infections in North America, China and Eastern Europe. The risk of being infected with HIV from a single prick with a needle that has been used on an HIV-infected person is thought to be about 1 in 150. Post-exposure prophylaxis with anti-HIV drugs can further reduce this risk.

This route can also affect people who give and receive tattoos and piercings. Universal precautions are frequently not followed in both sub-Saharan Africa and much of Asia because of both a shortage of supplies and inadequate training. The WHO estimates that approximately 2.5% of all HIV infections in sub-Saharan Africa are transmitted through unsafe healthcare injections. Because of this, the United Nations General Assembly has urged the nations of the world to implement precautions to prevent HIV transmission by health workers.

The risk of transmitting HIV to blood transfusion recipients is extremely low in developed countries where improved donor selection and HIV screening is performed. However, according to the WHO, the overwhelming majority of the world's population does not have access to safe blood and between 5% and 10% of the world's HIV infections come from transfusion of infected blood and blood products.

Perinatal Transmission

The transmission of the virus from the mother to the

child can occur in utero during the last weeks of pregnancy and at childbirth. In the absence of treatment, the transmission rate between a mother and her child during pregnancy, labour and delivery is 25%. However, when the mother takes antiretroviral therapy and gives birth by caesarean section, the rate of transmission is just 1%.

The risk of infection is influenced by the viral load of the mother at birth, with the higher the viral load, the higher the risk. Breastfeeding also increases the risk of transmission by about 4 per cent.

AIDS Denialism

A small group of activists, including several scientists who do not study HIV/AIDS, question the connection between HIV and AIDS, the existence of HIV itself, or the validity of current testing and treatment methods. Though these claims have been examined and thoroughly rejected by the scientific community, they continue to be promulgated through the Internet and have had a significant political impact, particularly in South Africa, where President Thabo Mbeki's embrace of AIDS denialism has been blamed for an ineffective response to that country's AIDS epidemic.

Misconceptions

A number of misconceptions have arisen surrounding HIV/AIDS. Three of the most common are that AIDS can spread through casual contact, that sexual intercourse with a virgin will cure AIDS and that HIV can infect only homosexual men and drug users. Other misconceptions are that any act of anal intercourse between gay men can lead to AIDS infection.

PATHOPHYSIOLOGY

The pathophysiology of AIDS is complex, as is the case with all syndromes. Ultimately, HIV causes AIDS by depleting $CD4^+$ T helper lymphocytes. This weakens the immune system and allows opportunistic infections. T lymphocytes are essential to the immune response and without them, the body cannot fight infections or kill cancerous cells. The mechanism of $CD4^+$ T cell depletion differs in the acute and chronic phases.

During the acute phase, HIV-induced cell lysis and killing of infected cells by cytotoxic T cells accounts for $CD4^+$ T cell depletion, although apoptosis may also be a factor. During the chronic phase, the consequences of generalized immune activation coupled with the gradual loss of the ability of the immune system to generate new T cells appear to account for the slow decline in $CD4^+$ T cell numbers.

Although the symptoms of immune deficiency characteristic of AIDS do not appear for years after a person is infected, the bulk of $CD4^+$ T cell loss occurs during the first weeks of infection, especially in the intestinal mucosa, which harbors the majority of the lymphocytes found in the body. The reason for the preferential loss of mucosal $CD4^+$ T cells is that a majority of mucosal $CD4^+$ T cells express the CCR5 coreceptor, whereas a small fraction of $CD4^-$ T cells in the bloodstream do so. HIV seeks out and destroys CCR5 expressing $CD4^+$ cells during acute infection. A vigorous immune response eventually controls the infection and initiates the clinically latent phase. However, $CD4^+$ T cells in mucosal tissues remain depleted throughout the infection, although enough remain to initially ward off life-threatening infections.

Continuous HIV replication results in a state of generalized immune activation persisting throughout the chronic phase. Immune activation, which is reflected by the increased activation state of immune cells and release of proinflammatory cytokines, results from the activity of several HIV gene products and the immune response to ongoing HIV replication. Another cause is the breakdown of the immune surveillance system of the mucosal barrier caused by the depletion of mucosal $CD4^+$ T cells during the acute phase of disease. This results in the systemic exposure of the immune system to microbial components of the gut's normal flora, which in a healthy person is kept in check by the mucosal immune system.

The activation and proliferation of T cells that results from immune activation provides fresh targets for HIV infection. However, direct killing by HIV alone cannot account for the observed depletion of $CD4^+$ T cells since only 0.01-0.10% of $CD4^+$ T cells in the blood are infected.

A major cause of $CD4^+$ T cell loss appears to result from their heightened susceptibility to apoptosis when the immune system remains activated. Although new T cells are continuously produced by the thymus to replace the ones lost, the regenerative capacity of the thymus is slowly destroyed by direct infection of its thymocytes by HIV. Eventually, the minimal number of $CD4^+$ T cells necessary to maintain a sufficient immune response is lost, leading to AIDS

Cells affected

The virus, entering through which ever route, acts primarily on the following cells:

- Lymphoreticular system:
  - CD_4+ T-Helper cells
  - CD_4+ Macrophages
  - CD_4+ Monocytes
  - B-lymphocytes
- Certain endothelial cells
- Central nervous system:
  - Microglia of the nervous system
  - Astrocytes
  - Oligodendrocytes
  - Neurones - indirectly by the action of cytokines and the gp-120

The Effect

The virus has cytopathic effects but how it does it is still not quite clear. It can remain inactive in these cells for long periods, though. This effect is hypothesized to be due to the CD_4-gp120 interaction.

- The most prominent effect of the HIV virus is its T-helper cell suppression and lysis. The cell is simply killed off or deranged to the point of being function-less (they do not respond to foreign antigens). The infected B-cells can not produce enough antibodies either. Thus the immune system collapses leading to the familiar AIDS complications, like infections and neoplasms (vide supra).

- Infection of the cells of the CNS cause acute aseptic meningitis, subacute encephalitis, vacuolar myelopathy and peripheral neuropathy. Later it leads to even AIDS dementia complex.
- The CD_4-gp120 interaction (vide supra) is also permissive to other viruses like Cytomegalovirus, Hepatitis virus, Herpes simplex virus, etc. These viruses lead to further cell damage i.e. cytopathy.

DIAGNOSIS

The diagnosis of AIDS in a person infected with HIV is based on the presence of certain signs or symptoms.

Since June 5, 1981, many definitions have been developed for epidemiological surveillance such as the Bangui definition and the 1994 expanded World Health Organization AIDS case definition. However, clinical staging of patients was not an intended use for these systems as they are neither sensitive, nor specific.

In developing countries, the World Health Organiza-tion staging system for HIV infection and disease, using clinical and laboratory data, is used and in developed countries, the Centers for Disease Control (CDC) Classification System is used.

WHO Disease Staging System

In 1990, the World Health Organization (WHO) grouped these infections and conditions together by introducing a staging system for patients infected with HIV-1. An update took place in September 2005. Most of these conditions are opportunistic infections that are easily treatable in healthy people.

- *Stage I:* HIV infection is asymptomatic and not categorized as AIDS
- *Stage II:* Includes minor mucocutaneous manifestations and recurrent upper respiratory tract infections
- *Stage III:* Includes unexplained chronic diarrhea for

longer than a month, severe bacterial infections and pulmonary tuberculosis

- *Stage IV:* Includes toxoplasmosis of the brain, candidiasis of the esophagus, trachea, bronchi or lungs and Kaposi's sarcoma; these diseases are indicators of AIDS.

CDC Classification System

There are two main definitions for AIDS, both produced by the Centers for Disease Control and Prevention (CDC). The older definition is to referring to AIDS using the diseases that were associated with it, for example, lymphadenopathy, the disease after which the discoverers of HIV originally named the virus.

In 1993, the CDC expanded their definition of AIDS to include all HIV positive people with a $CD4^+$ T cell count below 200 per μL of blood or 14% of all lymphocytes. The majority of new AIDS cases in developed countries use either this definition or the pre-1993 CDC definition. The AIDS diagnosis still stands even if, after treatment, the $CD4^+$ T cell count rises to above 200 per μL of blood or other AIDS-defining illnesses are cured.

HIV test

Many people are unaware that they are infected with HIV. Less than 1% of the sexually active urban population in Africa has been tested and this proportion is even lower in rural populations. Furthermore, only 0.5% of pregnant women attending urban health facilities are counseled, tested or receive their test results. Again, this proportion is even lower in rural health facilities. Therefore, donor blood and blood products used in medicine and medical research are screened for HIV.

HIV tests are usually performed on venous blood. Many laboratories use fourth generation screening tests which detect anti-HIV antibody (IgG and IgM) and the HIV p24 antigen. The detection of HIV antibody or antigen in a patient previously known to be negative is evidence of HIV infection. Individuals whose first specimen indicates evidence of HIV infection will have a repeat test on a second blood sample to confirm the results. The window period (the time between initial infection and the development of detectable antibodies against the infection) can vary since it can take 3–6 months to seroconvert and to test positive.

Detection of the virus using polymerase chain reaction (PCR) during the window period is possible and evidence suggests that an infection may often be detected earlier than when using a fourth generation EIA screening test.

Positive results obtained by PCR are confirmed by antibody tests. Routinely used HIV tests for infection in neonates, born to HIV-positive mothers, have no value because of the presence of maternal antibody to HIV in the child's blood. HIV infection can only be diagnosed by PCR, testing for HIV pro-viral DNA in the children's lymphocytes.

Prevention

Estimated per act risk for acquisitionof HIV by exposure route

Exposure Route	*Estimated infectionsper 10,000 exposuresto an infected source*
Blood Transfusion	9,000
Childbirth	2,500
Needle-sharing injection drug use	67
Percutaneous needle stick	30
Receptive anal intercourse	50
Insertive anal intercourse	6.5
Receptive penile-vaginal intercourse	10
Insertive penile-vaginal intercourse	5
Receptive oral intercourse	1
Insertive oral intercourse	0.5

assuming no condom use source refers to oral intercourseperformed on a man

The three main transmission routes of HIV are sexual

contact, exposure to infected body fluids or tissues and from mother to fetus or child during perinatal period. It is possible to find HIV in the saliva, tears and urine of infected individuals, but there are no recorded cases of infection by these secretions and the risk of infection is negligible.

Sexual Contact

The majority of HIV infections are acquired through unprotected sexual relations between partners, one of whom has HIV. The primary mode of HIV infection worldwide is through sexual contact between members of the opposite sex.

During a sexual act, only male or female condoms can reduce the chances of infection with HIV and other STDs and the chances of becoming pregnant. The best evidence to date indicates that typical condom use reduces the risk of heterosexual HIV transmission by approximately 80% over the long-term, though the benefit is likely to be higher if condoms are used correctly on every occasion.

The male latex condom, if used correctly without oil-based lubricants, is the single most effective available technology to reduce the sexual transmission of HIV and other sexually transmitted infections. Manufacturers recommend that oil-based lubricants such as petroleum jelly, butter and lard not be used with latex condoms, because they dissolve the latex, making the condoms porous. If necessary, manufacturers recommend using water-based lubricants. Oil-based lubricants can however be used with polyurethane condoms.

The female condom is an alternative to the male condom and is made from polyurethane, which allows it to be used in the presence of oil-based lubricants. They are larger than male condoms and have a stiffened ring-shaped opening and are designed to be inserted into the vagina.

The female condom contains an inner ring, which keeps the condom in place inside the vagina – inserting the female condom requires squeezing this ring. However, at present availability of female condoms is very low and the price remains prohibitive for many women.

Preliminary studies suggest that, where female condoms

are available, overall protected sexual acts increase relative to unprotected sexual acts, making them an important HIV prevention strategy. Studies on couples where one partner is infected show that with consistent condom use, HIV infection rates for the uninfected partner are below 1% per year.

Prevention strategies are well-known in developed countries, however, recent epidemiological and behavioral studies in Europe and North America have suggested that a substantial minority of young people continue to engage in high-risk practices and that despite HIV/AIDS knowledge, young people underestimate their own risk of becoming infected with HIV.

Randomized controlled trials have shown that male circumcision lowers the risk of HIV infection among heterosexual men by up to 60%. It is expected that this procedure will be actively promoted in many of the countries affected by HIV, although doing so will involve confronting a number of practical, cultural and attitudinal issues. Some experts fear that a lower perception of vulnerability among circumcised men may result in more sexual risk-taking behaviour, thus negating its preventive effects.

Exposure to Infected Body Fluids

Health care workers can reduce exposure to HIV by employing precautions to reduce the risk of exposure to contaminated blood. These precautions include barriers such as gloves, masks, protective eyeware or shields and gowns or aprons which prevent exposure of the skin or mucous membranes to blood borne pathogens.

Frequent and thorough washing of the skin immediately after being contaminated with blood or other bodily fluids can reduce the chance of infection. Finally, sharp objects like needles, scalpels and glass, are carefully disposed of to prevent needlestick injuries with contaminated items.

Since intravenous drug use is an important factor in HIV transmission in developed countries, harm reduction strategies such as needle-exchange programmes are used in attempts to reduce the infections caused by drug abuse.

Mother-to-child Transmission (MTCT)

Current recommendations state that when replacement feeding is acceptable, feasible, affordable, sustainable and safe, HIV-infected mothers should avoid breast-feeding their infant. However, if this is not the case, exclusive breast-feeding is recommended during the first months of life and discontinued as soon as possible.

Treatment

There is currently no vaccine or cure for HIV or AIDS. The only known methods of prevention are based on avoiding exposure to the virus or, failing that, an antiretroviral treatment directly after a highly significant exposure, called post-exposure prophylaxis (PEP). PEP has a very demanding four week schedule of dosage. It also has very unpleasant side effects including diarrhea, malaise, nausea and fatigue.

Antiviral Therapy

Current treatment for HIV infection consists of highly active antiretroviral therapy, or HAART. This has been highly beneficial to many HIV-infected individuals since its introduction in 1996 when the protease inhibitor-based HAART initially became available. Current optimal HAART options consist of combinations (or "cocktails") consisting of at least three drugs belonging to at least two types, or "classes," of antiretroviral agents.

Typical regimens consist of two nucleoside analogue reverse transcriptase inhibitors (NARTIs or NRTIs) plus either a protease inhibitor or a non-nucleoside reverse transcriptase inhibitor (NNRTI). Because HIV disease progression in children is more rapid than in adults and laboratory parameters are less predictive of risk for disease progression, particularly for young infants, treatment recommendations are more aggressive for children than for adults. In developed countries where HAART is available, doctors assess the viral load, rapidity in CD4 decline and patient readiness while deciding when to recommend initiating treatment.

HAART allows the stabilization of the patient's

symptoms and viremia, but it neither cures the patient of HIV, nor alleviates the symptoms and high levels of HIV-1, often HAART resistant, return once treatment is stopped.

Moreover, it would take more than the lifetime of an individual to be cleared of HIV infection using HAART. Despite this, many HIV-infected individuals have experienced remarkable improvements in their general health and quality of life, which has led to the plummeting of HIV-associated morbidity and mortality. In the absence of HAART, progression from HIV infection to AIDS occurs at a median of between nine to ten years and the median survival time after developing AIDS is only 9.2 months. HAART is thought to increase survival time by between 4 and 12 years.

For some patients, which can be more than fifty per cent of patients, HAART achieves far less than optimal results, due to medication intolerance/side effects, prior ineffective antiretroviral therapy and infection with a drug-resistant strain of HIV. Non-adherence and non-persistence with therapy are the major reasons why some people do not benefit from HAART. The reasons for non-adherence and non-persistence are varied. Major psychosocial issues include poor access to medical care, inadequate social supports, psychiatric disease and drug abuse.

HAART regimens can also be complex and thus hard to follow, with large numbers of pills taken frequently. Side effects can also deter people from persisting with HAART, these include lipodystrophy, dyslipidaemia, diarrhoea, insulin resistance, an increase in cardiovascular risks and birth defects. Anti-retroviral drugs are expensive and the majority of the world's infected individuals do not have access to medications and treatments for HIV and AIDS

Future Research

It has been postulated that only a vaccine can halt the pandemic because a vaccine would possibly cost less, thus being affordable for developing countries and would not require daily treatments. However, even after almost 30 years of research, HIV-1 remains a difficult target for a vaccine.

Research to improve current treatments includes decreasing side effects of current drugs, further simplifying drug regimens to improve adherence and determining the best sequence of regimens to manage drug resistance. A number of studies have shown that measures to prevent opportunistic infections can be beneficial when treating patients with HIV infection or AIDS.

Vaccination against hepatitis A and B is advised for patients who are not infected with these viruses and are at risk of becoming infected. Patients with substantial immunosuppression are also advised to receive prophylactic therapy for Pneumocystis jiroveci pneumonia (PCP) and many patients may benefit from prophylactic therapy for toxoplasmosis and Cryptococcus meningitis as well.

Alternative Medicine

Various forms of alternative medicine have been used to treat symptoms or alter the course of the disease. Acupuncture has been used to alleviate some symptoms, such peripheral neuropathy, but cannot cure the HIV infection. Several randomized clinical trials testing the effect of herbal medicines have shown that there is no evidence that these herbs have any effect on the progression of the disease, but may instead produce serious side-effects.

Some data suggest that multivitamin and mineral supplements might reduce HIV disease progression in adults, although there is no conclusive evidence on if they reduce mortality among people with good nutritional status. Vitamin A supplementation in children probably has some benefit. Daily doses of selenium can suppress HIV viral burden with an associated improvement of the CD4 count. Selenium can be used as an adjunct therapy to standard antiviral treatments, but cannot itself reduce mortality and morbidity.

Current studies indicate that that alternative medicine therapies have little effect on the mortality or morbidity of the disease, but may improve the quality of life of individuals afflicted with AIDS. The psychological benefits of these therapies are the most important use.

EPIDEMIOLOGY

The AIDS pandemic can also be seen as several epidemics of separate subtypes; the major factors in its spread are sexual transmission and vertical transmission from mother to child at birth and through breast milk. Despite recent, improved access to antiretroviral treatment and care in many regions of the world, the AIDS pandemic claimed an estimated 2.1 million (range 1.9–2.4 million) lives in 2007 of which an estimated 330,000 were children under 15 years. Globally, an estimated 33.2 million people lived with HIV in 2007, including 2.5 million children. An estimated 2.5 million (range 1.8–4.1 million) people were newly infected in 2007, including 420,000 children.

Sub-Saharan Africa remains by far the worst affected region. In 2007 it contained an estimated 68% of all people living with AIDS and 76% of all AIDS deaths, with 1.7 million new infections bringing the number of people living with HIV to 22.5 million and with 11.4 million AIDS orphans living in the region. Unlike other regions, most people living with HIV in sub-Saharan Africa in 2007 (61%) were women. Adult prevalence in 2007 was an estimated 5.0% and AIDS continued to be the single largest cause of mortality in this region. South Africa has the largest population of HIV patients in the world, followed by Nigeria and India.

South & South East Asia are second worst affected; in 2007 this region contained an estimated 18% of all people living with AIDS and an estimated 300,000 deaths from AIDS. India has an estimated 2.5 million infections and an estimated adult prevalence of 0.36%. Life expectancy has fallen dramatically in the worst-affected countries; for example, in 2006 it was estimated that it had dropped from 65 to 35 years in Botswana.

PROGNOSIS

Without treatment, the net median survival time after infection with HIV is estimated to be 9 to 11 years, depending on the HIV subtype and the median survival rate after diagnosis of AIDS in resource-limited settings where treatment is not available ranges between 6 and 19 months, depending on the study. In areas where it is widely available, the development of HAART as effective therapy for HIV infection and AIDS reduced the death rate from this disease by 80% and raised the life expectancy for a newly-diagnosed HIV-infected person to about 20 years.

As new treatments continue to be developed and because HIV continues to evolve resistance to treatments, estimates of survival time are likely to continue to change. Without antiretroviral therapy, death normally occurs within a year. Most patients die from opportunistic infections or malignancies associated with the progressive failure of the immune system. The rate of clinical disease progression varies widely between individuals and has been shown to be affected by many factors such as host susceptibility and immune function health care and co-infections, as well as which particular strain of the virus is involved.

AIDS was first reported June 5, 1981, when the U.S. Centers for Disease Control and Prevention recorded a cluster of Pneumocystis carinii pneumonia (now still classified as PCP but known to be caused by Pneumocystis jirovecii) in five homosexual men in Los Angeles. In the beginning, the Centers for Disease Control and Prevention (CDC) did not have an official name for the disease, often referring to it by way of the diseases that were associated with it, for example, lymphadenopathy, the disease after which the discoverers of HIV originally named the virus.

They also used Kaposi's Sarcoma and Opportunistic Infections, the name by which a task force had been set up in 1981. In the general press, the term GRID, which stood for Gay-related immune deficiency, had been coined. However, after determining that AIDS was not isolated to the homosexual community, the term GRID became misleading and AIDS was introduced at a meeting in July 1982. By September 1982 the CDC started using the name AIDS and properly defined the illness. A more controversial theory known as the OPV AIDS hypothesis suggests that the AIDS epidemic was inadvertently started in the late 1950s in the Belgian Congo by Hilary Koprowski's research into a poliomyelitis vaccine. According to scientific consensus, this scenario is not supported by the available evidence.

SOCIETY AND CULTURE

Stigma

AIDS stigma exists around the world in a variety of ways, including ostracism, rejection, discrimination and avoidance of HIV infected people; compulsory HIV testing without prior consent or protection of confidentiality; violence against HIV infected individuals or people who are perceived to be infected with HIV; and the quarantine of HIV infected individuals. Stigma-related violence or the fear of violence prevents many people from seeking HIV testing, returning for their results, or securing treatment, possibly turning what could be a manageable chronic illness into a death sentence and perpetuating the spread of HIV.

AIDS stigma has been further divided into the following three categories:

- *Instrumental AIDS stigma:* A reflection of the fear and apprehension that are likely to be associated with any deadly and transmissible illness.
- *Symbolic AIDS stigma:* The use of HIV/AIDS to express attitudes toward the social groups or lifestyles perceived to be associated with the disease.
- *Courtesy AIDS stigma:* Stigmatization of people connected to the issue of HIV/AIDS or HIV- positive people.

Often, AIDS stigma is expressed in conjunction with one or more other stigmas, particularly those associated with homosexuality, bisexuality, promiscuity and intravenous drug use.

In many developed countries, there is an association between AIDS and homosexuality or bisexuality and this association is correlated with higher levels of sexual prejudice such as anti-homosexual attitudes. There is also a perceived association between AIDS and all male-male sexual behaviour, including sex between uninfected men.

Economic Impact

HIV and AIDS affects economic growth by reducing the availability of human capital. Without proper nutrition, health care and medicine that is available in developed countries, large numbers of people are falling victim to AIDS. They will not only be unable to work, but will also require significant medical care.

The forecast is that this will likely cause a collapse of economies and societies in countries with a significant AIDS populationi. In some heavily infected areas, the epidemic has left behind many orphans cared for by elderly grandparents.

The increased mortality in this region will result in a smaller skilled population and labour force. This smaller labour force will be predominantly young people, with reduced knowledge and work experience leading to reduced productivity.

An increase in workers' time off to look after sick family members or for sick leave will also lower productivity. Increased mortality will also weaken the mechanisms that generate human capital and investment in people, through loss of income and the death of parents. By killing off mainly young adults, AIDS seriously weakens the taxable population, reducing the resources available for public expenditures such as education and health services not related to AIDS resulting in increasing pressure for the state's finances and slower growth of the economy.

This results in a slower growth of the tax base, an effect that will be reinforced if there are growing expenditures on treating the sick, training (to replace sick workers), sick pay and caring for AIDS orphans. This is especially true if the sharp increase in adult mortality shifts the responsibility and blame from the family to the government in caring for these orphans.

On the level of the household, AIDS results in both the loss of income and increased spending on healthcare by the household. The income effects of this lead to spending reduction as well as a substitution effect away from education and towards healthcare and funeral spending. A study in Côte d'Ivoire showed that households with an HIV/AIDS patient spent twice as much on medical expenses as other households.

NIACIN

Nicotinic acid was synthesized in 1873 but was left on the shelf as an organic compound unrelated to the severe pellagra afflictions occurring throughout the world at that time. Sixty years later the compound was shown to be present in coenzymes I and II and two years thereafter Elvehjem cured 'black tongue' in dogs with the vitamin. Niacin was postulated to be part of factor H for fish in 1937, but deficiency symptoms were not adequately described until reported in trout in 1947.

Nicotinic Acid

Niacin is the preferred nomenclature. Nicotinic acid amide or niacinamide is the common form in which the vitamin is physiologically active. Niacin is a white, crystalline solid, soluble in water and alcohol.

It is stable in the dry state and may be autoclaved for short periods without destruction. It is also stable to heat in mineral acids and alkali. Niacin is both a carboxylic acid and an amine and forms quaternary ammonium compounds because of its basic nature.

Acidic characteristics include salt formation with alkali. Niacin can be esterified easily, then converted to amides. Niacinamide is a crystalline powder soluble in water and ethanol and the dry material is stable up to about 60°C. In aqueous solutions it is stable for a short period when autoclaved. It is the form in which the vitamin is normally found in niacinamide adenine dinucleotide (NAD) and in niacinamide adenine dinucleotide phosphate (NADP).

POSITIVE FUNCTIONS

The major function of niacin in NAD and NADP is hydrogen transport in intermediary metabolism. Most of these enzyme systems function by alternating between the oxidized and reduced state of the coenzymes NAD-NADH and NADP-NADPH. Oxidation-reduction reactions may be anaerobic as when pyruvate acts as the hydrogen acceptor and lactate is formed, or the reactions may be coupled to electron transport systems with oxygen as the ultimate hydrogen acceptor; such aerobic reactions occur in respiration.

Both NAD and NADP are involved in the synthesis of high energy phosphate bonds which furnish energy for certain steps in glycolysis, in pyruvate metabolism, amino acid and protein metabolism and in photosynthesis. An inter-relationship between thiamine and niacin exists since both vitamins are coenzymes in intermediary metabolism whereby food is oxidized to furnish energy for physiological functions, to maintain homeostasis and for body temperature maintenance in homeotherms.

DEFICIENCY SYNDROME

The deficiency signs in fish niacin are more slowly exhausted under experimental conditions than are some of the other vitamins resulting in less defined and more slowly developing symptoms.

Niacin deficiencies in fish were experimentally induced in the late forties and early fifties by using basal diets which had a low niacin content. Loss of appetite and poor food conversion were the first signs noted. Then the fish turned dark and went off feed, followed by the appearance of lesions in the colon, erratic motion, oedema of the stomach and colon and muscle spasms while fish were apparently resting. A predisposition to sunburn in fish confined in the open in shallow ponds or raceways was described.

Common carp showed a congestion of the skin with subcutaneous hemorrhages. Common symptoms of niacin deficiency in most fish studied were muscular weakness and spasms, coupled with poor growth and poor food conversion.

Requirements

In homeotherms on a balanced test ration the niacin requirement is generally estimated to be about ten times that of the thiamine requirement. These rations generally contain considerable carbohydrate material to furnish energy to maintain body temperature. In fish the requirement appears to be twenty to thirty times that of the thiamine needs determined for the same test conditions and test rations. This difference may be due to low carbohydrate content of young fish diets and the higher protein content of these rations. Conversion of tryptophan to niacin occurs in the mammalian liver and possibly also in the liver of fish.

This conversion may account for the slow development of the niacin-deficiency syndrome in fish. However, after 10-14 weeks on diets devoid of niacin, deficiency symptoms did occur in several species of fish. The symptoms were reduced by replacement of niacin in the ration even when high protein diets containing an excess of tryptophan were fed. Too much niacin inhibits growth.

Sources and Protection

Niacin is found in most animal and plant tissues. Rich sources are yeast, liver, kidney, heart, legumes and green vegetables. Wheat contains more niacin than corn and the vitamin is also found in milk and egg products. The vitamin is very stable since it is generally found in coenzyme form in raw materials. Niacin added to the diet as a supplement remains relatively unaltered during diet manufacture, processing and storage.

Antimetabolites and Inactivation

Pyridine-3-sulphonic acid and 3-acetylpyridine are compounds structurally related to niacin and are antimetabolites for this vitamin in animals and in micro-organisms. Additional niacin can overcome the anti-metabolic effect. Deficiency symptoms in rats may be induced by 6-amino niacinamide. The symptoms are reversed by addition of ten times more niacinamide than the antimetabolite. Thioacetamide has been reported to be a niacin antagonist in fish.

Clinical Assessment

Salmon feeding actively in the oceans showed liver niacin content of 70-80ωg of niacin/g of fresh tissue. About half of this amount is present in fingerling salmon raised in fresh water at 12-15°C and fed test rations containing 40-50 per cent protein and niacin supplements 500-750 mg/kg dry diet. Urinary metabolities of niacin have been measured in other animals on standard niacin load in test rations containing a standard tryptophan load.

The technique is well developed to measure the N1-methyl derivative in mammalian urine. These data have not been reported for fish but metabolism chambers are available for collecting branchial and urinary wastes from large fish intubated with different diet material.

BIOTIN

Certain foods, notably liver and kidney, were found to be protective against a form of dermatitis resulting from the consumption of egg white by rats, in the early thirties. It was soon learned that a specific component of egg albumin, avidin, rendered dietary biotin unavailable, hence producing the symptoms. Biotin was variously called coenzyme R and 'vitamin H'. It was isolated by Du Vigneaud in 1941 and synthesized by workers of Merck and Company in 1943. Biotin was once thought to be part of factor H for fish. Blue slime patch disease due to biotin deficiency was reported in trout.

Biotin

Biotin is a monocarboxylic acid slightly soluble in water and alcohol and insoluble in organic solvents. Salts of the acid are soluble in water. Aqueous solutions or the dry material are stable at 100 C and to light. The vitamin is destroyed by acids and alkalis and by oxidizing agents such as peroxides or permanganate. Biocytin is a bound form of biotin isolated from yeast, plant and animal tissues.

Other bound forms of the vitamin can generally be liberated by peptic digestion. Oxybiotin has partial vitamin activity but oxybiotin sulphonic acid and other analogues are antimetabolites inhibiting the growth of bacteria.

Avidin, a protein found in raw egg white, binds biotin and makes it unavailable to fish and other animals. Heating to denature the protein makes the bound biotin available again to the fish. Biocytin or -biotinyl lysine (the epsilon amino group of lysine and the carboxyl of biotin being combined in a peptide bond) is hydrolyzed by the enzyme biotinase making the protein-bound biotin available.

POSITIVE FUNCTIONS

Biotin is required in several specific carboxylation and decarboxylation reactions, including the carboxylation of pyruvic acid to form oxaloacetic acid. It is part of the coenzyme of several carboxylating enzymes fixing CO_2 such as propionyl coenzyme A involved in the conversion of propionic acid to succinic acid in methylmalonyl coenzyme A. Biotin is also involved in the conversion of acetyl CO_2 to malonyl coenzyme A in the formation of long chain fatty acids.

It has possible involvement in citrulline synthesis and may have effects on purine and pyrimidine synthesis. It is involved in the conversion of unsaturated fatty acids to the stable cis form in the synthesis of biologically active fatty acids.

DEFICIENCY SYNDROME

Some signs of biotin deficiency in salmonids are skin disorders, muscle atrophy, lesions in the colon, loss of appetite and spastic convulsions. Haematology discloses fragmentation of erythrocytes.

Poor growth is a common symptom and has been reported for salmonids, common carp, goldfish (Carassius auratus) and eel. Blue slime patch disease in brook trout deficient in biotin appears typical for this species. Fish reared in 10 -15 C water exhaust biotin stores in 8-12 weeks and the first signs are anorexia, poor food conversion and general listlessness before the more acute deficiency symptoms become detectable.

Sources and Protection

Rich sources of biotin are liver, kidney, yeast, milk products and egg yolks. Nut meats contain good supplies of biotin. The diet should be protected from strong oxidizing agents or conditions which promote oxidation of ingredients. Raw egg white should not be incorporated into moist fish diets. Cooking will inactivate the avidin.

Antimetabolites and Inactivation

Raw egg white has already been discussed which irreversibly binds biotin and makes it unavailable to young fish. Many biotin homologues with different side chain lengths inhibit the growth of bacteria. Oxybiotin, in which sulphur is replaced by oxygen, has about the same biological activity as the natural biotin. On the other hand, oxybiotin sulphonic acid inhibits biotin activity.

Clinical Assessment

Measurement or urinary excretion of biotin in animals does not prove a good clinical method since biotin is synthesized by several organisms in the gut. In fish, this technique needs to be explored and may well be valuable as a clinical tool since the gastrointestinal tract of many freshwater fish contains only small quantities of bacteria. Biotin is one of the most expensive vitamins to add to fish rations. Salmon feeding actively in the oceans had liver biotin concentrations of 10-12 mg of wet liver tissue.

The concentrations in the liver of young salmon fingerlings in fresh water fed test diets containing an excess of biotin were between 6-8 mg of the vitamin/g of tissue. Fish with these levels of biotin in the liver should probably be in sound biotin nutritional status.

FOLIC ACID (FOLACIN)

A factor effective in curing metaloblastic anaemia in monkeys was found in yeast and liver concentrates and was designated vitamin M in 1935. In 1939 an anti-anaemic factor was found in liver and called vitamin BC. These were later shown to be the same substance with folic acid as the active ingredient. Folic acid was synthesized in 1946 and was soon used in fish diets for preventing purified diet anaemia.

Folic Acid

Folic acid crystallizes into yellow spear-shaped leaflets which are soluble in water and dilute alcohol. It can be precipitated with heavy metal salts.

It is stable to heat in neutral or alkaline solution, but unstable in acid solution. It deteriorates when exposed to sunlight, or during prolonged storage. Several analogues have biological activity including pteroic acid, rhizopterin, folinic acid, xanthopterin and several formyltetrahydropteroylglutamic acid derivatives.

These have closely allied ring structures and many have been isolated as derivatives in various animals or microbiological preparations. One simple form, xanthopterin, present in the pigments of insects, is shown below and is of special interest because of early work with this compound as the anti-anaemic factor H for fish.

Positive Functions

Folic acid is required for normal blood cell formation and is involved as a coenzyme in one-carbon transfer mechanisms.

In the presence of ascorbic acid, folic acid is transformed into the active (5-formyl-5, 6,7,8) tetrahydrofolic acid. Folic acid is involved in many one-carbon metabolism systems such as serine and glycine interconversion, methionine-homocysteine synthesis, histidine synthesis and pyrimidine synthesis.

Several coenzyme forms of the active vitamin have been isolated. Folic acid is involved in the conversion of megaloblastic bone marrow to normoblastic type. It has a role in blood glucose regulation and improves cell membrane function and hatchability of eggs.

Deficiency Syndrome

Macrocytic normochromic anaemia occurs in several experimental animals, including fish fed diets devoid of the folic acid. Increasing numbers of senile cells are observed as the deficiency progresses until only a few old and degenerating cells are found in the blood of deficient fish. Anterior kidney imprints disclose only adult cells and no preforms present. Other signs observed have been poor growth, anorexia, general anaemia, lethargy, fragile fins, dark skin pigmentation and infarction of spleen.

SOURCES AND PROTECTION

Yeast, green vegetables, liver, kidney, glandular tissue, fish tissue and fish viscera are good sources of folic acid. Insects contain xanthopterin which has folic acid activity. At one time the yellow pigment of xanthopterin was identified as the fish anti-anaemic factor H, but subsequent experiments showed only partial activity and that folic acid itself was a much more potent antimacrocytic anaemia factor.

Insects may contribute significantly to the folic acid requirements of wild fish, but in scientific fish husbandry artificial diets are more reliable sources. Activity is lost during extended storage and when material is exposed to sunlight. Therefore, dry feeds should be carefully protected during manufacture and moist diet rations should be carefully preserved. Both types of fish diets should be fed soon after manufacture to assure minimal loss of folic acid activity.

ANTIMETABOLITES AND INACTIVATION

One antagonist of folacin is 4-aminopteroylglutamic acid or aminopterin. This material, when incorporated in the diet of guinea pigs and rats, induces anaemia and leucopenia and has been used to treat leukemia in man. Amethopterin (4-amino-N10-methylpteroylglutamic acid) can also be used to induce deficiency by inhibiting purine, pyrimidine and nucleic acid protection; viz., inhibition of nucleic acid synthesis results in macroytic anaemia.

CLINICAL ASSESSMENT

Haematology is used as a simple clinical tool to assess haemopoiesis in fish. Anterior kidney imprints easily disclose normal distribution of immature cells and preforms undergoing reticulosis. Salmon actively feeding in the ocean and young salmon fingerlings raised in fresh water and fed diets rich in folacin show liver storage of 3-4ωg of folic acid/g of wet tissue. Microbiological assay is preferred for assessment of total folic acid in dietary raw materials because the total biological activity it measures includes all the various coenzyme forms and folic acid analogues.

Assessment of the dietary intake of folic acid is important for intensive cold water fish husbandry. In pond culture, aquatic and terrestrial insects, algae, etc., may also be available and folic acid in the supplementary feed may not be as critical. Since folic acid is labile in storage, excess amounts are generally added to manufactured feed in anticipation of storage losses.

However, prudent fish husbandry dictates rapid use of manufactured rations with minimum storage. Routine periodic haematology of fish assures proper nutritional status for maximum production and sound health. The author has noted in several series of experiments that when fish diseases occur through inadvertent contamination of the water supply, those groups of fish partially or completely deficient in folic acid were among the first to show acute disease symptoms. Therefore, folic acid must also play an important role in resistance to disease.

VITAMIN B12

The antipernicious anaemia factor found to be contained in liver was isolated and crystallized in the mid forties. This substance, named vitamin B12 by its discoverers, was later to be recognized as essential for growth of chicken fed diets entirely of plant origin and was designated animal protein factor (APF). When anaemic salmon were injected with crystalline B12 in combination with folic acid and xanthopterin positive haemopoiesis occurred within a few days and the salmon showed rapid recovery from the anaemia.

CHEMISTRY

Vitamin B12 or cynacobalamin has the following approximate chemical structure.

Cyanocobalamin

The molecule has a planar group and a nucleotide group lying nearly at right angles to one another. This cobalt-containing vitamin has a net charge of one at the central cobalt atom to which is attached a replaceable cyano group. Vitamin

B12 is stable to mild heat in neutral solution, but is rapidly destroyed by heating in dilute acid or alkali. Crude concentrates are more unstable and rapidly lose activity.

The compound is similar to the porphyrins in its spatial configuration with a central cobalt atom linked to four reduced pyrrole rings in the haeme series. Replacing the cyanide ion with a variety of anions produce derivatives which have comparable biological activities; viz., hydroxocobalamin, nitritocobalamin, chlorocobalamin and sulphatocobalamin.

Positive Functions

Cyanocobalamin is involved with folic acid in haemopoiesis. It is required by many micro-organisms and is a growth factor for many animals. The animal protein factor present in fish and animal by-products was not recognized until crystalline vitamin B12 was injected into anaemic chinook salmon fingerlings in 1949 and positive haemopoiesis was observed. A coenzyme incorporating vitamin B12 is involved in the reversible isomerization of methyl-malonyl coenzyme A to succinyl coenzyme A and in the isomerization of methylaspartate to glutomate.

Cyanocobalamin is involved in the coenzyme for the methylation of homocystine to form methionine. It is also involved in several other one-carbon reactions and in the synthesis of labile methyl compounds. One vitamin B12 containing coenzyme acts in methylation of the purine ring during thymine synthesis. Vitamin B12 is also involved in cholesterol metabolism, in purine and pyrimidine biosynthesis and in the metabolism of glycols.

Deficiency Syndrome

Deficiency signs in young pigs, chicks and rats show abnormal blood elements, poor growth and pernicious anaemia. An intrinsic factor is necessary for good absorption of the vitamin from the gut. This factor is a low molecular weight mucoprotein which normally occurs in gastric juice and especially in hog gut mucosa. Pernicious anaemia in chinook and coho salmon is characterized by fragmented, erythrocytes with many aberrant forms present.

Haemoglobin levels are inconsistent and erythrocyte counts have a range extending from frank anaemia to a near normal blood pattern. Cyanocobalamin stores in fish tissues are slowly exhausted and only after 12-16 weeks on test do the symptoms appear in deficient salmon populations. Poor appetite, poor growth, poor food conversion and some dark pigmentation can be observed before frank anaemia is detected.

Sources and Protection

Rich sources of vitamin B12 are found in fish meal, fish viscera, liver, kidney, glandular tissues and slaughter house wastes. Since vitamin B12 is labile on storage and in mild acid solution is easily destroyed by heating, care must be exercised in diet preparation containing flesh or meat scraps.

Antimetabolites and Inactivation

The vitamin B12 coenzymes are very unstable under light, which rapidly decomposes the coenzymes. Photo-sensitivity is increased in dilute acid solutions.

Clinical Assessment

Generalized anaemia with fragmentation of erythrocytes and extremely variable haemoglobin levels and erythrocyte counts indicate possible B12 deficiency. Prompt response in individual fish is obtained by injecting B12 alone or in combination with folic acid in the ratio of 1 part vitamin B12 to 100 parts folic acid. Careful interpretation of haemotological data will enable one to distinguish one form of anaemia from the other.

ASCORBIC ACID

Experimental work to cure scurvy with fruit juice was described by Lind in 1753, but nearly 200 years elapsed before the exact chemical compound responsible for reducing the symptoms was defined. The isolation of ascorbic acid was due to Szent-Györgyi of Hungary and C.G. King of the U.S.A. Vitamin C synthesis was accomplished in 1933 after the chemical structure of ascorbic acid was established by British and Swiss workers.

McCay and Tunison reported scoliosis in brook trout fed

formalin-preserved meat in 1934 and McLaren observe haemorrhages in trout fed rations low in ascorbic acid. It was not until the sixties that a critical need for L-ascorbic acid by trout and salmon was demonstrated. L-ascorbic acid is a white, odourless, crystalline compound, soluble in water but insoluble in fat solvents.

It is readily oxidized to dehydroascorbic acid, the less biologically potent form. Ascorbic acid is very stable in acid solution because of the preservation of the lactone ring, but in alkaline solution hydrolysis occurs rapidly and vitamin activity is lost.

It is very heat labile and prone to atmospheric oxidation, especially in the presence of copper, iron, or several other metallic catalysts. The reduced form is the most biologically active but several derivatives or salts are obtainable which have varying degrees of ascorbate activity.

Positive Functions

L-ascorbic acid acts as a biological reducing agent in hydrogen transport. It is involved in many enzyme systems for hydroxylation; i.e., hydroxylation of tryptophan, tyrosine, or proline.

It is involved in the detoxification of aromatic drugs and also acts in the production of adrenal steroids. Ascorbic acid is necessary for the formation of hydroxy proline which is a constituent of collagen, a component of intercellular material in bones and soft tissues.

Ascorbic acid plays a synergistic role with vitamin E as intracellular antioxidants and free radical traps. The conversion of folic acid to folinic acid requires vitamin C. Ascorbic acid is involved in the formation of chondroitin sulphate and intercellular ground substance.

Labelled ascorbic acid fed to fish previously deficient in the vitamin was shown to be rapidly mobilized and fixed in areas of rapid collagen synthesis and became concentrated in the thick collagen of the skin and in cartilagenous bones, as well as in the glands of the anterior kidney. Ascorbic acid is also involved in erythrocyte maturation.

Deficiency Syndrome

Scurvy with impaired collagen formation, perifollicular haemorrhages, loose teeth and poor bone formation, anaemia and oedema have been reported in other animals. Deficiency signs in fish are generally related to impaired collagen formation.

Fish soon show hyperplasia of jaw and snout. The same symptoms have been observed in trout, salmon, yellowtail, carp, guppies and char. Histologically, hypertrophy of the adrenal tissues and haemorrhage at the bases of fins have been observed in coho salmon. Deficiency signs cease to develop and new growth becomes normal upon replacement of ascorbic acid in the ration. Anaemia eventually develops in extremely deficient fish and scoliosis and lordosis do not repair but are walled off by new growth around the afflicted areas of the spine when ascorbic acid is once again added to the ration.

Requirements

In the rainbow trout, reasonable blood and anterior kidney storage levels were obtained with an intake of about 100 mg of vitamin C/kg of dry ration in 10°, 12°, or 15° C water systems. When wound repair experiments were initiated, however, or when fish were exposed to other stress then the requirements doubled or tripled.

When severe abdominal or intramuscular wounds were inflicted, young fish needed at least 500 mg of active ascorbate for tissue repair comparable with control fish receiving 1 g or more of ascorbate in the diet/kg of dry diet. Coho salmon appear to need about half of these requirements for adequate tissue levels and for maximum severe wound repair rates.

The requirement for ascorbic acid is related to stress, growth rate and size of the animal, as well as to the other nutrients present in the diet.

A compromise value of about 200 mg of ascorbic acid/kg diet for trout and salmon raised in freshwater systems between 10-15°C would ensure reasonable tissue storage levels and furnish some excess for mild stress conditions and for ascorbic acid loss from the diet through oxidation. Large common carp can synthesize some ascorbate and the requirement for this species may be dependent on fish size and the environment in which they are reared.

Table. Growth and Tissue Ascorbate

Vitamin C diet treatment	Trout			Salmon		
	Average weight at 24 weeks	*Ascorbate concentrate* [1]		*Average weight at 24 weeks*	*Ascorbate concentrate* [1]	
		blood	*kidney*		*blood*	*kidney*
mg/100 g	*g*	*g/g*	*g/g*	*g*	*g/g*	*s/s*
0	2.4	[2]	[2]	5.0	22.3±2.2	89
5	9.6	34.4±2.9	125	6.0	30.5±1.2	132
10	10.6	34.6±1.3	137	5.7	35.8±1.6	265
20	10.1	38.8±3.3	132	6.1	34.2±2.3	183
40	10.2	46.8±6.2	162	6.3	33.7±2.0	225
100	10.8	51.0±4.6	247	6.0	37.8±2.3	321

1. Average of five samples for blood (± S.D.) and two for head kidney tissue
2. No fish available for assay

Sources and Protection

Ascorbic acid is widely distributed in nature with citrus fruits, cabbage, liver and kidney tissue good sources for the vitamin.

Fresh insects and fish tissues contain reasonable amounts of the vitamin. Synthetic ascorbic acid is also readily available. Fish food should be protected from oxidizing agents and kept sealed or frozen until used to prevent loss of the vitamin.

Antimetabolites and Inactivation

D-ascorbic acid, the optical isomer of the active form, has no vitamin activity and competes for sites of several enzyme reactions mediated by L-ascorbic acid. 6-deoxy-L-ascorbic acid, dehydroascorbic acid and L-glucoascorbic acid have very little activity.

Clinical Assessment

Ascorbic acid status for experimental animals is normally

attempted by tissue ascorbate analysis. Most of the assays previously used measure total ascorbate and not biologically active L-ascorbic acid. Consequently these are fraught with errors and misconception of true vitamin C status. In fish tissues, blood and liver do not adequately reflect the ascorbic acid intake and status, whereas assay of the anterior kidney which contains adrenal tissue provides a fairly representative picture of tissue storage of the vitamin.

Stress rapidly reduces the ascorbic acid content of this tissue with concurrent production of adrenal steroids. Conversely, dietary repletion is reflected by up to four or five-fold storage levels from the deficient state. Examination of fragile support cartilage in the gill filaments under low magnification will detect early hypovitaminosis before clinically acute symptoms become noticeable.

However, the best tissue for routine clinical analysis to assess vitamin C status in trout and salmon appears to be the anterior kidney with samples selected from the junction of the two forward wings. Tissues are blotted free of blood with filter paper and then assayed for total ascorbate by one of the improved quick methods to determine total ascorbate.

INOSITOL

Muscle 'sugar' was discovered by Scherer in 1850 and was characterized as inositol in 1887. Inositol was shown to be an alopecia (a form of hairlessness) preventing factor for mice in 1940. Poor growth and poor food passage in inositol-deficient fish was observed in salmon and carp.

Inositol

Seven optically inactive and two optically active isomers of hexahydroxycyclohexane can exist. One of the optically active forms, myo-inositol, is a white crystalline powder soluble in water and insoluble in alcohol and ether. The material can be synthesized, but is easily isolated from biological material in free or combined forms. The mixed calcium-magnesium salt of the hexophosphate is phytin. Isomers have little biological activity but will compete in chemical reactions. Inositol is a highly stable compound.

POSITIVE FUNCTIONS

Myo-inositol is a structural component in living tissues. It has lipotropic action by preventing accumulation of cholesterol in one type of fatty liver disease and is involved with choline in maintaining normal lipid/metabolism. It is a growth-promoting substance for micro-organisms and prevents an alopecia in mice. It is a reserve carbohydrate in muscle as well as a major component of phosphoglycerides in animal tissues.

DEFICIENCY SYNDROME

Poor growth, increased gastric emptying time, oedema, dark colour and distended stomachs are symptoms observed in salmon, trout, carp and catfish held for long periods on inositol-deficient test rations. A 'spectacle-eye' condition described for rats has not been observed under the experimental conditions used in fish studies. The major deficiency sign is inefficiency in digestion and food utilization and concomitant poor growth leading to a population of fish with distended abdomens.

SOURCES AND PROTECTION

Myo-inositol occurs ubiquitously in large amounts wherever biological tissue is found. Wheat germ, dried peas and beans are rich sources. Brain, heart and glandular tissues are very good sources of biologically active inositol. Citrus fruit pulp and dried yeast also contain inositol. The compound is stable.

ANTIMETABOLITES AND INACTIVATION

Seven optically inactive and one optically active but biologically inactive stereo isomers occur. Since inositol is synthesized in the biologically active form by many microorganisms in the gut, only large quantities of chemically synthesized inactive isomers added to diets would interfere with the metabolism of inositol for growth and normal physiological function. Biologically inactive stereoisomers of myo-inositol do not compete for critical sites in metabolism. Methyl derivatives and mono-, di- and triphosphoric acid esters occur naturally. Salts of the hexaphosphate or phytin make the bound inositol practically unavailable to the animal.

CLINICAL ASSESSMENT

Assessment in fish has been based on lack of deficiency signs coupled with the most efficient food conversion. Salmon feeding actively in oceans show 1-1.5 mg of inositol/g of fresh liver tissue and young fingerlings raised in freshwater at 10-15°C had 600-700ωg/g of liver tissue. An alternate, better assessment may be based on a standard muscle section or whole carcass analysis for free or for bound inositol. Projection of inositol intake from normal fish diet ingredients should indicate an excess of this particular vitamin.

CHOLINE

Methylation as a basic metabolic process was postulated by Hoffmeister. Methyl transfer was shown in vivo by Thompson and the interrelationships between choline, methionine and homo-cystine were shown by du Vigneaud in 1939-42. Trout fed low choline rations developed haemorrhagic kidneys and salmon showed an aversion to food in choline-deficient diet.

Choline

Choline is a very strong organic base and forms many derivatives widely distributed in animal and vegetable tissue. One derivative, acetylcholine, is involved in the transmission of nerve impulses across synapses. Choline is very hygroscopic, very soluble in water and is stable to heat in acid, but decomposes in alkaline solutions.

Positive Functions

Choline acts as a methyl donor in trans-methylation reactions. It is a lipotropic and antihaemorrhagic factor preventing the development of fatty livers. It is involved in the synthesis of phospholipids and in fat transport.

Acetylcholine transmits the excitory state across the ganglionic synapses and neuromuscular junctions. Choline is essential for growth and good food conversion in fish.

Deficiency Syndrome

Deficiency signs include poor growth and poor food conversion accompanied by impaired fat metabolism. Haemorrhagic kidneys and intestines have been reported in trout and increased gastric emptying time has been observed in salmon.

Sources and Protection

Rich sources of choline are wheat germ, beans, brain and heart tissue. Choline hydrochloride, the commercially available form, may inactivate -tocopherol and vitamin K when in direct contact with these vitamins and care should be exercised in selecting properly protected (gelatin coated) fat soluble vitamins in fish diet preparation.

Clinical Assessment

Choline status of fish can be estimated from the assay of choline content of the dietary ingredients with the absence of deficiency signs.

Maximum liver storage may not be the best criteria to determine choline nutritional status but has been used to assess the tentative requirement listed for the two species of salmon.

p-AMINOBENZOIC ACID

Para-aminobenzoic acid is a white crystalline powder which is water soluble and heat and light stable in aqueous and mild alkaline solution.

It is a growth promoting vitamin for micro-organisms which require the vitamin for folic acid synthesis. Large intake has been shown to counteract the antimetabolite effect of sulphonamides in bacterial culture. No positive function or deficiency signs have been observed in fish.

LIPOIC ACID

Lipoic acid is both fat soluble and water soluble. Its active form is the amide derivative known as lipoamide. Lipoic acid functions as a coenzyme in a -ketoacid decarboxylation. It was discovered independently in several laboratories during the period 1945-50 and shown to be an essential component of multienzyme, the pyruvate dehydrogenase complex. Pyruvate is converted to 'active acetaldehyde which in turn is picked up by lipoamide. Subsequent oxidation of the aldehyde results in the reduction of lipoamide to a disulphydryl form. The multi-enzyme unit also includes thiamine pyrophosphate, coenzyme A and flavin adenine dinucleotide. Glandular tissues are good sources of lipoic acid. No requirements have been determined for fish.

FAT-SOLUBLE VITAMINS

Fat-soluble vitamins A, D, E and K differ from the water-soluble vitamins in their accumulative action. Little evidence has been recorded for hypervitaminosis with the water-soluble vitamins since these compounds are rapidly metabolized and excreted when intake exceeds liver or tissue storage capacity, but hypervitaminosis is a common occurrence in fish and other animals when large quantities of any one of the fat-soluble vitamins are ingested.

Toxicity symptoms involving vitamins A and D are indistinguishable from deficiency symptoms for the same vitamins.

On the other hand, symptoms of excess vitamins E and K intake are more discrete. Fish rations may often be enriched with fish oils to increase caloric density of the ration resulting in excessive intake of the fat-soluble vitamins.

VITAMIN A

A fat-soluble factor that promoted growth in rats was described by Hopkins and by Osborne and Mendel at the beginning of this century. McCollum and Simmonds cured xerophthalmia, a disease of the eye, with this material. The chemical structure of vitamin A and its relationship to -

carotene was shown by Von Euler in 1928. Active vitamin A was synthesized in the mid-thirties.

The relationship of the vitamin A alcohols to naturally occurring -carotene is as follows:

Retinene, the aldehyde form of vitamin A, has been isolated from the retina of dark adapted eyes and is involved in vision in dim light. Retinoic acid, which is the oxidized form of vitamin A alcohol, has been shown to have some vitamin A activity. Vitamin A1 is found in saltwater fish, whereas vitamin A is more abundant in freshwater fish.

Interconversion of the two forms in living fish tissue has been demonstrated. Fish oils contain vitamin A as free alcohols or esters. Vitamin A alcohol occurs as a light coloured viscous oil which is heat labile and subject to air oxidation.

Beta-carotene occurs as an orange, crystalline compound which is more stable to heat and oxidation. Vitamin A is water insoluble but is soluble in fat and organic solvent.

POSITIVE FUNCTIONS

Vitamin A is essential in maintaining epithelial cells, preventing atrophy and keratinization of epithelial tissues. It is combined with a protein in visual purple and is important in night vision. Vitamin A also prevents xerophthalmia in rats and young children. Vitamin A promotes growth of new cells and aids in maintaining resistance to infection.

THE SYNDROME OF DEFICIENCY OR EXCESS

Hypovitaminosis A is characterized by poor growth, poor vision, keratinization of epithelial tissue, xerophthalmia, night blindness, haemorrhage in the anterior chamber of the eye, haemorrhage of the base of the fins and abnormal bone formation. Nerve degeneration has been reported in pigs, chickens, rats, rabbits and ducks, but only occasionally observed in fish after long periods of deficiency.

Hypervitaminosis A has been described in fish and in other animals and involved enlargement of liver and spleen, abnormal growth, skin lesions, epithelial keratinization, hyperplasia of head cartilage and abnormal bone formation resulting in ankylosis and fusion of vertebrae. Hypervitaminosis A is reflected in very high liver oil vitamin A content and elevated serum alkaline phosphatase. Removal of excess vitamin A from the diet promotes rapid recovery.

SOURCES AND PROTECTION

Cod liver oil is the best known source of vitamin A although black sea bass, swordfish, or ling cod oils contain much more of the vitamin. Whale liver oil contains kitol which has little or no biological activity until heated above 200° C when one molecule of biologically active vitamin A is generated per molecule of whale kitol. This biologically inactive kitol may be deposited in the whale to avoid hypervitaminosis A during excessive vitamin A intake.

The possibility of hypervitaminosis A occurs when tuna, shark, or ling cod viscera are used in preparation of moist fish diets. Synthetic vitamin A preparations, such as vitamin A palmitate, are available and are often used to supplement rations low in fish meal, fish viscera, or carotenes. Some fish species seem able to utilize b-carotene as a vitamin A source, whereas others are unable to split the g-carotene molecule and vitamin A must be added to the diet.

CLINICAL ASSESSMENT

Vitamin A status can best be assessed by the absence of deficiency signs and by assay of liver oil for vitamin A content. Assay for the vitamin in blood or plasma has not been found to be useful.

VITAMIN D

Rickets was induced with test diets by Hopkins and Mellanby cured the disease in dogs by adding cod liver oil to the ration.

Ultraviolet light exposure was shown to have an anti-rachitic effect and provitamin D activity had been ascribed to ergosterol. Crystalline vitamin D was isolated by Angus and activated 7-dehydrocholesterol was isolated by Windaus.

Vitamin D_2 or ergocalciferol, is one of several biologically active forms of vitamin D. Vitamin D_3 or activated 7-dehydrocholesterol has the chemical formula $C_{27}H_{44}O$ and contains a more simplified, unsaturated 8-carbon side chain. Vitamin D_3, also known as cholecalciferol, is formed in most animal tissue by the rupture of one of the ring bonds of 7-dehydrocholesterol by ultraviolet radiation.

Cholecalciferol is a white, crystalline compound soluble in fat and organic solvents and is stable to heat and oxidation in mild alkali or acid solution.

POSITIVE FUNCTIONS

Vitamin D is essential for maintaining calcium and inorganic phosphate homeostasis. It is involved in alkaline phosphatase activity, promotes intestinal absorption of calcium and influences the action of parathyroid hormone on bone. Fish may sequester calcium from water through the gill membrane; thus the major function of vitamin D for other animals may not be necessary to satisfy calcium requirements for fish.

SYNDROME OF DEFICIENCY OR EXCESS

Hypovitaminosis D for fish has not been described. Rickets and abnormal bone formation described in detail for animals has not been observed in fish fed low vitamin D diets. However little work has been done under carefully controlled conditions with young growing fish and only alteration in alkaline phosphatase activity has been reported on different vitamin D diet intake. Hypervitaminosis D, however, has been reported. Brook trout fed large doses of vitamin D showed impaired growth, lethargy and dark colouration.

High intake of vitamin D mobilizes phosphorus and calcium from the bone and tissues and may result in fragile bones, poor growth and poor appetite related to the nausea described in man afflicted with hyper-vitaminosis D. Arterial and kidney lesions due to excessive intake of vitamin D reported for rats and dogs have not been described histologically for fish nor has hypercalcaemia been described.

This area needs to be explored because of the potential

for hypervitaminosis D in fish fed diets containing various fish viscera which might contain large amounts of the vitamin D. Tuna liver oil may contain, for example, 100 to 1 000 times as much active vitamin D as cod liver oil.

SOURCES AND PROTECTION

Vitamin D requirements of many animals can be met by ultraviolet irradiation of the skin where cholesterol derivatives are converted to calciferol. Since the vitamin is fat soluble and accumulates in lipid stores, fish liver oil is a rich source of the material.

Content varies tremendously in liver oil, however, with values of about 25 I.U./g present in shark liver oil and over 200 000 I.U./g in albacore tuna liver oil. Cod liver oil contains from 100-500 I.U./g and animal liver contains some vitamin D. One international unit (I.U.) is equal to 0.025 mg of crystalline vitamin D.

CLINICAL ASSESSMENT

Absorption maxima in the ultraviolet region can be used to detect provitamins D in the non-saponifiable fraction of oils. Concentrated preparations of vitamin D can be assayed by the Carr-Price antimony trichloride reaction when assay is necessary to determine biologically active materials in liver oil of fish on different treatments.

The chick assay may not apply to fish liver storage levels because the biologically active form for fish has not been determined. Therefore, clinical assessment for hypervitaminosis A must rely on crude methods for determination of vitamin D in liver oil samples by the Carr-Price reaction or by measuring absorption in the UV spectrum.

VITAMIN E

The existence of an antisterility vitamin was postulated by Evans and Bishop. This factor was named 'vitamin E' by Sure. The active tocopherol was isolated, characterized and synthesized by Karrer.

Vitamin E is composed of a class of compounds known

as tocopherols. One of the most important tocopherols is -tocopherol.

Eight naturally occurring tocopherol derivatives have been isolated and all belong to the D series. Synthetic -tocopherol is a racemic DL-tocoperhol mixture. The derivatives of tocol or of tocotrienol are named, alpha, beta, gamma, delta, epsilon, eta, zeta1 and zeta2-tocopherol The pure tocopherols are fat-soluble oils which are capable of esterification to form crystalline compounds.

The tocopherols are stable to heat and acids in the absence of oxygen, but are rapidly oxidized in the presence of nascent oxygen, peroxides, or other oxidizing agents. The tocopherols are sensitive to ultraviolet light and are excellent antioxidants in the free form.

The esters are more stable and are commonly used as dietary supplements-anticipating hydrolysis in the gut and absorption of the free alcohol to act as an active intra- and intercellular antioxidant. Ethyl derivatives on the aromatic ring are also active. Oxidation products of a-tocopherol can be reduced with hydrosulphite to -tocopherylhydroquinone or, in the presence of ascorbic acid, to -tocopherol.

Table. The Tocopherols

Tocopherol (alpha)	5,7,8-trimethyltocol
Tocopherol (beta)	5,8-dimethyltocol
Tocopherol (gamma)	7,8-dimethyltocol
Tocopherol (zeta$_2$)	5,7-dimethyltocol
Tocopherol (eta)	7-methyltocol
Tocopherol (delta)	8-methyltocol
Tocopherol (epsilon)	5,8-dimethyltocotrienol
Tocopherol (zeta$_1$)	5,7,8-trimethyltocotrienol

POSITIVE FUNCTIONS

The tocopherols act as extra and intracellular antioxidants to maintain homeostasis of labile metabolites in the cell and tissue plasma. As physiological antioxidants, these usually protect oxidizable vitamins and labile unsaturated fatty acids.

Vitamin E prevents encephalomalacia in chicks, erythrocyte haemolysis in several animals and steatitis in mink, pigs and farm animals. The tocopherols also prevent exudative diathesis in chicks, white muscle disease in lambs and calves and dietary liver necrosis in rats.

They act with selenium and with vitamin C for normal reproductive activity and are involved in prevention of nutritional muscular dystrophy in the chick, the yellowtail and carp. The tocopherols act as free radical traps to stop the chain reaction during peroxide formation and stabilize unsaturated carbon bonds of polyunsaturated fatty acids and other long-chain labile compounds.

Vitamin E is involved in the maintenance of normal blood capillary permeability and the integrity of heart muscle. It was first shown to be involved in prevention of sterility and foetal resorption in rats. It may likewise affect embryo membrane permeability and hatch-ability of fish eggs.

SYNDROME OF DEFICIENCY OR EXCESS

One of the first signs is erythrocyte fragility closely followed by anaemia, ascites, xerophthalmia, poor growth, poor food conversion, epicarditis and ceroid deposits in spleen and liver. Muscle dystrophy and xerophthalmia have been described in yellowtail and carp. Non-specific forms of degenerative conditions have been described in several species of fish fed large quantities of polyunsaturated fatty acids with inadequate tocopherol in the ration. Hypervitaminosis E results in poor growth, toxic liver reaction and death.

SOURCES AND PROTECTION

Vegetable oils are rich sources of tocopherols. Synthetic - tocopherol in esterfied acetate or phosphate form is commonly used as a diet supplement. These esters are much more stable than the free form which is rapidly lost by air oxidation or in the presence of labile compounds like the polyunsaturated fish oils. Addition of antioxidants such as BHT (butylated hydroxyanisole) and BHT (butylated hydroxytoluene) protects fats and other labile compounds in the ration from oxidation, but these antioxidants have no vitamin E activity.

CLINICAL ASSESSMENT

The erythrocyte fragility test indicates the physiological state of fish. Absence of histologically detectable ceroid in liver and spleen from representative samples in the population is a good clue on the presence of adequate amounts of physiological antioxidants in the fish.

A barbituric acid test for oxidation of components has not been applied to fish tissues as a clinical tool for assessment of nutrition state except when liver oils become saturated with very labile polyunsaturated fatty acids such as when feeding squid or saury oils. These assays are good indicators of the state of oxidation of the finished ration, but the absence of deficiency signs and normal erythrocyte fragility are better clinical tests. Analysis for tocopherol is difficult and time-consuming and only applicable under critical research experiment situation.

VITAMIN K

The name vitamin K (for 'koagulation') was proposed by Dam. Dam isolated the vitamin from alfalfa and from fish meal in 1939. It was synthesized later that year.

The K vitamins K_2 contain 6,7, or 9 isoprene units in the side chain which varies from 30-45 carbon atoms. Many isosteres of vitamin K have been identified in animal tissues, plant tissues and micro-organisms. The structure of phthiocol and menadione (K_3) are shown below:

Although fairly stable compounds, they are destroyed upon oxidation and exposure to ultraviolet radition. Menadione is very reactive and is subject in aqueous media to chemical interaction.

POSITIVE FUNCTIONS

Vitamin K is involved in the hepatic synthesis of blood clotting proteins - prothrombin and proconvertin.

Substituted forms of vitamin K are strongly bacteriostatic and may serve as an alternate defence mechanism for bacterial infections. Vitamin K is structurally related to flavo compounds which function as electron carriers in the electron transport system. The primary role of vitamin K is to maintain a fast normal blood clotting rate which is so important to fish living in a water environment.

SYNDROME OF DEFICIENCY OR EXCESS

Prothrombin time in salmon fed diets devoid of vitamin K was increased three to five times and, during prolonged deficiency states, anaemia and haemorrhagic areas appeared in the gills, eyes and vascular tissues. Increased blood clotting time has also been reported for other fish reared on diets with low vitamin K content.

Interrelationships with other vitamins have not been documented in fish where the primary deficiency signs are slow blood clotting and haemorrhage, severe anaemia, or death in wounded fish. Haemorrhagic areas often appear in fragile tissues such as the gills.

SOURCES AND PROTECTION

Vitamin K sources are green, leafy vegetables. Alfalfa leaves are among the best sources of vitamin K. Low levels are found in soybeans and animal liver. Synthetic menadione is also available. Although vitamin K in ground alfalfa is fairly stable, synthetic material should be protected from exposure to ultraviolet light and to excessive oxidizing or reducing conditions. The use of rapidly cured dehydrated alfalfa is essential to minimize formation of the physiological antagonist, dicumarol. The diet should be kept dry, prepared with minimum exposure to air oxidation and fed as soon as practicable after manufacture to minimize vitamin K loss.

Chapter 9

Uses of Microbes

The enslavement and death of countless millions of workers may result from a newly discovered method of producing glycerine (currently produced as a byproduct of soap manufacture). Whether the process reported by Canada's National Research Council in Ottawa will be adopted commercially depends largely on the efficiency of the workers' digestive apparatus. In any case, no one will protest their exploitation. The workers are microscopic members of the clan Bacillus subtilis.

The use of microbes as minuscule chemical factories has been practiced, if not understood, since the first butter was churned, the first wine pressed, the first beer brewed. Spurred by advances in the field of biochemistry and the pressures of two wars, the employment of microbe labour has recently spread to a whole new field of chemical manufacture. Little is known of the actual metabolic process by which microbes work, but by careful control and endless experiment chemists and biologists, working together, have been able to set them hundreds of chemical tasks.

Best known of these industrial microorganisms is yeast, whose appetite for the carbohydrates in beets, sugar cane, wood and other fibrous vegetable matter made possible the production in 1944 of about 638 million gallons of alcohol—grain and wood. But the yeasts are only one group of the microbic multitude able to perform specific jobs. Bacteria resembling the bacilli of human ailments and molds like mildew have also been put to work in industry.

Some microbes have done double jobs. One, fed in a

certain way, yields oxalic acid, basic chemical of the blueprint industry; on a different diet it produces the gluconic acid used in medicines. The versatile Clostridium acetobutylicum, on a single diet of corn mash, produces acetone for solvents, butanol for automobile lacquers and riboflavin (Vitamin 62).

Grasshoppers and Cantaloupes. In a constant search for new microscopic workers, industrial, university and Government researchers have isolated over 200,000 varieties. Useful microbes may turn up anywhere—in the air, on the water, on forest leaf mold, in city garbage cans. A potent industrial bacillus was discovered in the intestines of a grasshopper. The best strain of the mold Penicillium notatum, which makes life-saving penicillin, was first noticed on a cantaloupe rind. Despite their humble origins, microbes are as temperamental as coloraturas.

They are fussy about temperature, about food and about the company they keep. Some like plenty of air, some like none and some govern their behaviour according to whether they get it or not. In an airless, quiet place, yeast will produce wine; in air it just reproduces itself. To keep such un reliable workers healthy, happy and productive is the responsibility of a growing new species of industrial scientist, the biological engineer. His job: to reproduce on a factory scale biological processes and conditions none too easy to control in a laboratory test tube.

Biological Engineering. Leaders in the development of this field have been technicians of such firms as Brooklyn's Charles Pfizer & Co., one of the oldest of the fermentation chemical makers and a big producer of penicillin. They have devised techniques that smack less of a factory than of the contagion ward of a hospital.

The production of penicillin, which is typical of most fermentation processes, starts from a small culture of the mold growing on a mixture of water, sugar and some nitrogenous food. A 10,000-gal. vat of nutrient may be charged with as much as 100 gal. of this mixture, which has been kept sterilized so that no impurity can possibly be present. Should another, different microbe find his way in, the batch would be spoiled.

In theory, all that is necessary to start production is to introduce the microscopic microbe into the large vat. In practice, the vats are inoculated with the larger cultures, to save time, but at the end of three days the biological engineer may find the clear, brownish liquid filled with the clusters of strawlike gold that he had hoped to produce. Or he may find only an accumulation of horrible smells.

Microbe Mysteries. The biologist is often completely in the dark as to why his microbes fail so dismally, when they do. Cross contamination of one culture by another is a common cause. To prevent it at Pfizer's, no two chemicals are made in the same building. Workers with coughs or sneezes are kept away and those going in & out are thoroughly sterilized and dressed like hospital interns. Filtered air and ultra violet light are constantly at work.

In spite of all precautions, the microbes find ways to flout their keepers. A simon-pure culture may sometimes, apparently from sheer perversity, change its nature. After such a mutation, it may produce a better product, a worse product, or an entirely different product. Biologists are no more certain of the reason for this than they are of the nature of the sub-microscopic organisms which plague the microbes themselves. Absenteeism among microbe workers is no problem, but in at least one factory devoted to the manufacture of essential chemical solvents, all work had to be stopped for days because the microbes themselves got sick.

PRODUCTION AND USE OF MICROBIAL ENZYMES FOR DAIRY PROCESSING

Indians are known to be lovers of milk and its products. As a sequel to white revolution, India has surged ahead to become the largest milk producer in the world, the production figure touched to 74 million tons for the year 1998. With growing urbanization, demand for processed dairy foods has increased considerably, in particular demand for different cheese varieties and low-lactose milk due to increasing intolerance of human beings to lactose in milk and other milk products. For improving the quality of milk and milk products, a number of different enzymes from microbial as well as from nonmicrobial sources have potential applications in dairy processing.

The use of rennet in cheese manufacture was among the earliest applications of exogenous enzymes in food processing, dating back to approximately 6000 B C. In 1994, the total production of cheese was 8000 metric tons against a total demand of 9000 metric tons. The projected demand by 2000 A D is around 30,000 metric tons. The use of rennet, as an exogenous enzyme, in cheese manufacture is perhaps the largest single application of enzymes in food processing.

In recent years, proteinases have found additional applications in dairy technology, for example in acceleration of cheese ripening, modification of functional properties and preparation of dietic products. Principal among some enzymes that have important and growing applications are lipases and b-galactosidases.

Enzymes with limited applications include glucose oxidase, superoxide dismutase, sulphydryl oxidase, etc. The increasing use of enzymes to produce specific products with characteristic attributes can be emphasized by the world-wide sale of industrial enzymes approximating to US $ 1.6 billion, which is expected to reach US $ 3.0 billion by the year 2008.

Almost 45% of this is shared by the food industry and the remaining is shared by detergent (34.4%), textile (11%), leather (2.8%), pulp and paper (1.2%) industries and other industries (5.6%) excluding enzymes for use in diagnostics and therapeutics. India being the highest producer of milk in the world and consequently the surplus availability of milk in our country has triggered the food and dairy industry to convert the liquid milk into value-added products using biochemical and enzymatic processes.

Microbial Rennets in Dairy Applications

Animal rennet (bovine chymosin) is conventionally used as a milk-clotting agent in dairy industry for the manufacture of quality cheeses with good flavour and texture. Owing to an increase in demand for cheese production world wide –

i.e. 4% per annum over the past 20 years, approximating 13.533 million tons – coupled with reduced supply of calf rennet, has therefore led to a search for rennet substitutes, such as microbial rennets. At present, microbial rennet is used for one-third of all the cheese produced world wide.

Rennin acts on the milk protein in two stages, by enzymatic and by nonenzymatic action, resulting in coagulation of milk. In the enzymatic phase, the resultant milk becomes a gel due to the influence of calcium ions and the temperature used in the process. Many microorganisms are known to produce rennet-like proteinases which can substitute the calf rennet.

Microorganisms like Rhizomucor pusillus, R. miehei, Endothia parasitica, Aspergillus oryzae and Irpex lactis are used extensively for rennet production in cheese manufacture. Extensive research that has been carried out so far on rennet substitutes has been reviewed by several authors. Different strains of species of Mucor are often used for the production of microbial rennets.

Whereas best yields of the milk-clotting protease from Rhizomucor pusillus are obtained from semisolid cultures containing 50% wheat bran, R. miehei and Endothia parasitica are well suited for submerged cultivation. Using the former, good yields of milk-clotting protease may be obtained in a medium containing 4% potato starch, 3% soybean meal and 10% barley. During growth, lipase is secreted together with the protease. Therefore, the lipase activity has to be destroyed by reducing the pH, before the preparation can be used as cheese rennet.

Microbial rennets from various microorganisms (marketed under the trade names such as Rennilase, Fromase, Marzyme, Hanilase, etc.) being marketed since the 1970s have proved satisfactory for the production of different kinds of cheese. The molecular and enzymatic properties of chymosins have been studied extensively. Although the proteolytic specificities of the three commonly used fungal rennets are considerably different from those of calf chymosin, these rennets have been used to produce acceptable cheeses. Recently Novo Nordisk has succeeded in expressing just one proteolytic enzyme from the fungus R. miehei in the well-known organism A. oryzae.

This host organism is able to produce the single protease that cleaves the casein into a glycomacropeptide and para casein by hydrolysing only at the phe105–met106 peptide bond between phenyl alanine and methionine. This monocomponent enzyme product has the trade name Novoren. One major drawback of microbial rennet use in cheese manufacture, is the development of off flavour and bitter taste in the nonripened as well as in the ripened cheeses.

The rennets from microbial sources are more proteolytic in nature in comparison to rennet from animal sources, resulting in production of some bitter peptides during the process of cheese ripening. Hence, attempts have been made to clone the gene for calf chymosin and to express it in selected bacteria, yeasts and molds.

Recombinant Rennets for Cheese Manufacture

Due to shortage of calf stomachs and the economic value of cheese rennet, gene for calf chymosin was one of the first genes for mammalian enzymes that was cloned and expressed in microorganisms. Many different laboratories have cloned the gene for calf prochymosin in Escherichia coli and analysed the structure of the gene as well as the properties of the recombinant chymosin.

The expressed proenzyme in E. coli, is present mainly as insoluble inclusion bodies, comprised of reduced prochymosin as well as molecules that are interlinked by disulphide bridges. After disintegration of the cells, inclusion bodies are harvested by centrifugation.

The individual laboratories have reported some differences in the procedure for renaturation of prochymosin from the inclusion bodies, but all have followed the same general scheme. The enzymatic properties of recombinant E. coli chymosin are indistinguishable from those of native calf chymosin. The enzymes were identical when observed by immunodiffusion in gels, but a slight difference was observed by enzyme linked immunosorbent assay (ELISA).

Table. Recombinal Chymosin Proparations at/oı Approaching Legal and Commercial Acceptance.

Sourcc of DNA	*Praducing microorganisms*	*Prdι cing company (brand namc)*
Calf abomasums	klayveromyces lactis	Gist Brocades (Maxiren)
Calf abomasums	Aspergillus niger	Ger ccor/chr. Hansen (Chyrnoger)
Synthetic	Escherichic coli	Pfizer (Chymax)

The gene for prochymosin has also been cloned in Saccharomyces cerevisiae; the levels of expression have been reported to be 0.5 to 2.0% of total yeast protein. In yeast, about 20% of the prochymosin can be released in soluble form which can be activated directly; the remaining 80% is still associated with the cell debris.

The cloning of chymosin was carried out without the prosequence. Though the level of chymosin mRNA was similar to that of prochymosin mRNA, no milk-clotting activity was observed in clones containing the chymosin gene only. The results suggest that the prosequence is essential for correct folding of the polypeptide chain.

The zymogen for the aspartic proteases from R. pusillus, also called mucor rennin, has likewise been cloned and expressed in yeast. Studies on its conversion to active form showed that secretion of R. pusillus protease from recombinant yeast was dependent on glycosylation of the enzyme.

Compared to yeast, the filamentous fungi generally secrete larger quantities of proteins into the culture medium. Furthermore, filamentous fungi secrete the heterologous proteins with correct folding of the polypeptide chain and process correct pairing of sulphydryl groups.

The gene for R. miehei protease has been expressed in A. oryzae. Prochymosin has been expressed in Kluyveromyces lactis, A. nidulans, A. niger and Trichoderma reesei. In most of the cases, the reported yields of the model systems were about 10–40 mg of enzyme per litre of culture medium.

However, 3.3 g of enzyme per liter has also been achieved. Yeast, Kluyveromyces lactis, has recently been used as an efficient host for the secretion of recombinant chymosin, which has led to the development of large-scale production process for chymosin.

If produced on an industrial scale, the yields will perhaps be of the latter magnitude. Chymosins which are approaching or have marketing and legal acceptance and generally regarded as safe (GRAS).

Most of the companies produce recombinant rennet of cattle calf origin in different microbial hosts, however in India, the major source of milk is buffalo, showing a different composition from that of cow. The rennet from buffalo source has inherent compatibility for clotting buffalo milk.

Therefore, National Dairy Research Institute, Karnal, has taken a lead in cloning gene of buffalo chymosin. The buffalo chymosin cDNA has been cloned in E. coli and partial N-terminal sequencing of the purified buffalo chymosin indicates that it is highly homologous to the cattle chymosin.

Several cheese making experiments have been carried out with recombinant chymosin and the general aspects of recombinant chymosin have been dealt with in a report. Since most of the rennet (> 90 per cent) added to cheese milk is lost in the whey, immobilization would considerably extend its catalytic life. Several rennets have been immobilized, but their efficiency as milk coagulants has been questioned.

So, there is a fairly general support for the view that immobilized enzymes cannot coagulate milk properly owing to inaccessability of the Phe–Met peptide bond of k-casein and that the apparent coagulating activity of immobilized rennets is due to leaching of the enzyme from the support. Different types of conventional cheeses have been successfully made by using recombinant rennet on an experimental or pilot scale. No major differences have been detected between cheeses made with recombinant chymosins or natural enzymes, regarding cheese yield, texture, smell, taste and ripening. The recombinant chymosins are identical with calf rennet according to the report on biochemical and genetic evidences.

Lactase in Dairy Industry

Lactose, the sugar found in milk and whey and its corresponding hydrolase, lactase or b -galactosidase, have been extensively researched during the past decade This is because of the enzyme immobilization technique which has given new and interesting possibilities for the utilization of this sugar. Because of intestinal enzyme insufficiency, some individuals and even a population, show lactose intolerance and difficulty in consuming milk and dairy products.

Hence, low-lactose or lactose-free food aid programme is essential for lactose-intolerant people to prevent severe tissue dehydration, diarrhoea and, at times, even death. Another advantage of lactase-treated milk is the increased sweetness of the resultant milk, thereby avoiding the requirement for addition of sugars in the manufacture of flavoured milk drinks.

Manufacturers of ice cream, yoghurt and frozen desserts use lactase to improve scoop and creaminess, sweetness and digestibility and to reduce sandiness due to crystallization of lactose in concentrated preparations Cheese manufactured from hydrolysed milk ripens more quickly than the cheese manufactured from normal milk.

Technologically, lactose crystallizes easily which sets limits to certain processes in the dairy industry and the use of lactase to overcome this problem has not reached its fullest potential because of the associated high costs. Moreover, the main problem associated with discharging large quantities of cheese whey is that it pollutes the environment. But, the discharged whey could be exploited as an alternate cheap source of lactose for the production of lactic acid by fermentation. The whey permeate, which is a by-product in the manufacture of whey protein concentrates, by ultrafiltration could be fermented efficiently by Lactobacillus bulgaricus.

Lactose can be obtained from various sources like plants, animal organs, bacteria, yeasts (intracellular enzyme), or molds. Some of these sources are used for commercial enzyme preparations. Lactase preparations from A. niger, A. oryzae and Kluyveromyces lactis are considered safe because these sources already have a history of safe use and have been subjected to numerous safety tests. The most investigated E. coli lactase is not used in food processing because of its cost and toxicity problems.

Properties of Lactase

The properties of the enzyme depend on its source. Temperature and pH optima differ from source to source and with the type of particular commercial preparation. Immobilization of the enzymes, method of immobilization and type of carrier can also influence these optima values. In general, fungal lactase have pH optima in the acidic range 2.5–4.5 and yeast and bacterial lactases in the neutral region 6–7 and 6.5–7.5, respectively.

The variation in pH optima of lactases makes them suitable for specific applications, for example fungal lactases are used for acid whey hydrolysis, while yeast and bacterial lactases are suitable for milk (pH 6.6) and sweet whey (pH 6.1) hydrolysis. Product inhibition, e.g. inhibition by galactose, is another property which also depends on the source of lactase.

The enzyme from A. niger is more strongly inhibited by galactose than that from A. oryzae. This inhibition can be overcome by hydrolysing lactose at low concentrations by using immobilized enzyme systems or by recovering the enzyme using ultrafiltration after batch hydrolysis. Lactase from Bacillus species are superior with respect to thermostability, pH operation range, product inhibition and sensitivity against high-substrate concentration.

Thermostable enzymes, able to retain their activity at 60°C or above for prolonged periods, have two distinct advantages viz. they give higher conversion rate or shorter residence time for a given conversion rate and the process is less prone to microbial contamination due to higher operating temperature. Bacillus species have a pH optima of 6.8 and a temperature optima of 65°C. Its high activity for skim milk and less inhibition by galactose has made it suitable for use as a production organism for lactase.

The enzymatic hydrolysis of lactose can be achieved either by free enzymes, usually in batch fermentation process, or by immobilized enzymes or even by immobilized whole cells producing intracellular enzyme. Although numerous hydrolysis systems have been investigated, only few of them have been scaled up with success and even fewer have been applied at an industrial or semi-industrial level.

Several acid hydrolysis systems have been developed to industry-scale level. Large-scale systems which use free enzyme process have been developed for processing of UHT-milk and processing of whey, using K. lactis lactase (Maxilact, Lactozyme). Several commercial immobilized systems have been developed for commercial exploitation. Snamprogretti process of industrial-scale milk processing technology in Italy is one such working systems.

They make use of fibre-entrapped yeast lactase in a batch process and the milk used is previously sterilized by UHT. For pilot plants, there are three other processes designed and developed to handle milk:

- Gist-Brocades, Rohm GmbH (Germany);
- Sumitomo, Japan. These are continuous processes with short residence times. Processing of whey UF-permeate is accomplished by the system developed by Corning Glass, Connecticut, Lehigh, Valio and Amerace corp. The process by Corning Glass is being applied at commercial scale in the bakers yeast production using hydrolysed whey.

Microbial Enzymes in Accelerated Cheese Ripening

Cheese ripening is a complex process mediated by biochemical and biophysical changes during which a bland curd is developed into a mature cheese with characteristic flavour, texture and aroma. The desirable attributes are produced by the partial and gradual breakdown of carbohydrates, lipids and proteins during ripening, mediated by several agents, viz.

- Residual coagulants,
- Starter bacteria and their enzymes,

- Nonstarter bacteria and their enzymes,
- Indegenous milk enzymes, especially proteinases;
- Secondary inocula with their enzymes.

Proteolysis occurs in all the cheese varieties and is a prerequisite for characteristic flavour development that can be regulated by proper use of the above agents. Cheese ripening is essentially an enzymatic process which can be accelerated by augmenting activity of the key enzymes.

This has the advantage of initiating more specific action for flavour development compared to use of elevated temperatures that can result in accelerating undesirable nonspecific reactions and consequently off flavour development.

Enzymes may be added to develop specific flavours in cheeses, for example lipase addition for the development of Parmesan or Blue-type cheese flavours. Attempts to accelerate the multiple secondary flavour-forming-reactions, e.g. Strecker degradation, have been scarce. The pathways leading to the formation of flavour compounds are largely unknown and therefore the use of exogenous enzymes to accelerate ripening is mostly an empirical process. Different microbial enzymes used to accelerate cheese ripening.

Studies on the hydrolysis of whole casein and isolated casein components to observe the kinetics and specificity of aspartic proteases in rennin, pepsin and four microbial rennet substitutes indicated the considerable differences in the reaction velocity and the extent of hydrolysis on the rennet curd yield. The rennet enzymes were active even in the later phases of cheese-ripening.

Proteinases and Peptidases

Proteolysis is characteristic of most cheese varieties and is indispensable for good flavour and textural development. Proteinases used in cheese processing include:

- Plasmin,
- Rennet;
- Proteinases (cell wall and/or intracellular) of the starter and nonstarter bacteria. Approximately 6% of the

rennet added to cheese milk remains in the curd after manufacture and contributes significantly to proteolysis during ripening. Combinations of individual neutral proteinases and microbial peptidases intensified cheese flavour and when used in combinations with microbial rennets reduced the intensity of bitterness caused by the latter. Acid proteases in isolation cause intense bitterness. Various animal or microbial lipases gave pronounced cheese flavour, low bitterness and strong rancidity, while lipases in combination with proteinases and/or peptidases give good cheese flavour with low levels of bitterness. In a more balanced approach to the acceleration of cheese ripening using mixtures of proteinases and peptidases, attenuated starter cells or cell-free extracts (CFE) are being favoured.

Table. Microbial Enzymes used to Ucoclcrate Chocse Ripening

Enzyme	*Source*
Microbial lipases	Aspergillus niger
	A. oryzae
	Rhizomucor miehei
Lactasce	Streptococcus lactis
	Ktuyveromyces sp.
	Escherichia coli
Microbial serline proteinases	A. niger
Neutral pmroteinases	Bacillus subtilis A.oryzae

PROTEOLYTIC ENZYMES OF LACTIC ACID BACTERIA IN FERMENTED MILK PRODUCTS

The proteolytic system of lactic acid bacteria is essential for their growth in milk and contributes significantly to flavour development in fermented milk products. The proteolytic system is composed of proteinases which initially cleaves the milk protein to peptides; peptidases which cleave the peptides to small peptides and amino acids; and transport system responsible for cellular uptake of small peptides and amino acids.

Lactic acid bacteria have a complex proteolytic system capable of converting milk casein to the free amino acids and peptides necessary for their growth. These proteinases include extracellular proteinases, endopeptidases, aminopeptidases, tripeptidases and proline-specific peptidases, which are all serine proteases. Apart from lactic streptococcal proteinases, several other proteinases from nonlactostreptococcal origin have been reported.

There are also serine type of proteinases, e.g. proteinases from Lactobacillus acidophillus, L. plantarum, L. delbrueckii sp. bulgaricus, L. lactis and L. helveticus. Aminopeptidases are important for the development of flavour in fermented milk products, since they are capable of releasing single amino acid residues from oligopeptides formed by extracellular proteinase activity.

Other Dairy Enzymes

Other enzymes used for dairy food application include:

- Proteases to reduce allergic properties of cow milk products for infants;
- Lipases for development of lipolytic flavours in speciality cheeses.

The functional properties of milk proteins may be improved by limited proteolysis through the enzymatic modification of milk proteins. An acid-soluble casein, free of off flavour and suitable for incorporation into beverages and other acid foods, has been prepared by limited proteolysis. The antigenicity of casein is destroyed by proteolysis and the hydrolysate is suitable for use in milk-protein-based foods for infants allergic to cow milk.

Lipolysis makes an important contribution to swiss cheese flavours, due mainly to the lipolytic enzymes of the starter cultures. The characteristic peppery flavour of Blue cheese is due to short-chain fatty acids and methyl ketones. Most of the lipolysis in Blue cheese is catalysed by Penicillium roqueforti lipase, with a lesser contribution from indigenous milk lipase. The NOVO process for production of enzyme modified cheese (EMC) uses medium-aged cheese which is emulsified, homogenized and pasteurized, after which 'palatase' (a lipase from R. miehei) is added, with or without a proteinase and the blend is ripened at a high temperature for one to four days.

The mixture is reheated, a paste results which is suitable for inclusion in soups, dips, dressings, or snack foods. EMC technology has been developed to produce a range of characteristic cheese flavours and flavour intensities, for example swiss, blue, cheddar, provolo-nemor or romano, suitable for inclusion at low levels in many products.

The claims that exogenous enzymes are effective in accelerating ripening have not led to their wide-spread use, possibly due to their high cost, difficulties in distributing them uniformly in the curd and the possible danger of over-ripening the cheese. The other minor enzymes having limited applications in dairy processing include glucose oxidase, catalase, superoxide dismutase, sulphydryl oxidase, lactoperoxidase and lysozymes.

Glucose oxidase and catalase are often used together in selected foods for preservation. Superoxide dismutase is an antioxidant for foods and generates H_2O_2, but is more effective when catalase is present. Thermally induced generation of volatile sulphydryl groups is thought to be responsible for the cooked off-flavour in ultra high temperature (UHT) processed milk.

Use of sulphydryl oxidase under aseptic conditions can eliminate this defect. The natural inhibitory mechanism in raw milk is due to the presence of low levels of lactoperoxidase (LP), which can be activated by the external addition of traces of H_2O_2 and thiocyanate.

It has been reported that the potential of LP-system and its activation enhances the keeping quality of milk. Cow milk can be provided with protective factors by the addition of lysozyme, making it suitable as an infant milk.

Lysozyme acts as a preservative by reducing bacterial counts in milk without affecting the The scope of application of minor enzymes to milk and milk products has been recently reviewed.

The global market for the production of microbial enzymes for use in dairy-products manufacture is considerably large, but is being dominated only by a limited number of enzyme producers.

In India the microbial dairy enzymes requirement has been very limited till now. However, with the advent of technological processes for the manufacture of different varieties of milk products, such as cheeses by the State Dairy Federations, Co-operatives and Private Dairy Product Manufacturers like Amul, Vijaya, Verka, Dynamix, Nestle, Smith Kline Beechem, etc., the markets for the sale of such products in megacities and towns has been slowly growing for the past two to three years.

Presently, many of these microbial enzymes, such as microbial rennets and other enzymes, are being imported. Hence, there is a scope for the production of enzymes such as microbial rennet, lactase, proteinases and lipases indigenously. In the near future, the requirement for these enzymes is bound to increase by leaps and bounds, basically due to requirement of value-added dairy products in the country.

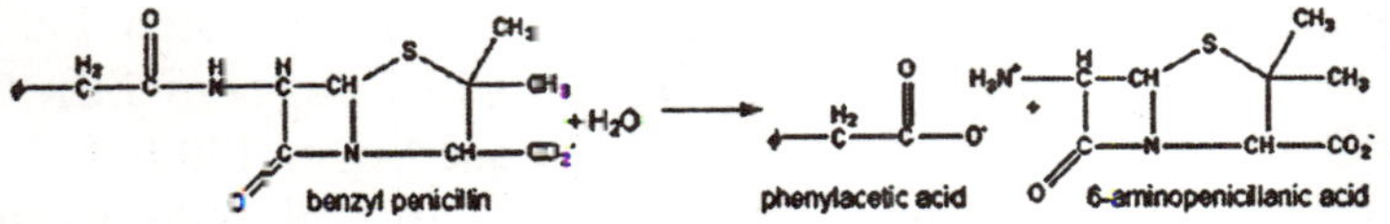

benzyl penicillin $+ H_2O \rightarrow$ phenylacetic acid + 6-aminopenicillanic acid

PRODUCTION OF ANTIBIOTICS

The mass production of antibiotics began during World War II with streptomycin and penicillin. Now most antibiotics are produced by staged fermentations in which strains of microorganisms producing high yields are grown under optimum conditions in nutrient media in fermentation tanks holding several thousand gallons.

The mold is strained out of the fermentation broth and then the antibiotic is removed from the broth by filtration, precipitation and other separation methods. In some cases new antibiotics are laboratory synthesized. while many antibiotics are produced by chemically modifying natural substances; many such derivatives are more effective than the natural substances against infecting organisms or are better absorbed by the body, e.g., some semisynthetic penicillins are effective against bacteria resistant to the parent substance.

Benzylpenicillins and phenoxymethylpenicillins (penicillins G and V, respectively) are produced by fermentation and are the basic precursors of a wide range of semi-synthetic antibiotics, e.g. ampicillin.

The amide link may be hydrolysed conventionally but the conditions necessary for its specific hydrolysis, whilst causing no hydrolysis of the intrinsically more labile but pharmacologically essential b-lactam ring, are difficult to attain. Such specific hydrolysis may be simply achieved by use of penicillin amidases (also called penicillin acylases).

Different enzyme preparations are generally used for the hydrolysis of the penicillins G and V, pencillin-V-amidase being much more specific than pencillin-G-amidase. Penicillin amidase may be obtained from E. coli and has been immobilised on a number of supports including cyanogen bromide-activated Sephadex G200.

It represents one of the earliest successful processes involving immobilised enzymes and is generally used in batch or semicontinuous STR processes (40,000 Ukgpenicillin G, 35°C, pH 7.8, 2 h) where it may be reused over 100 times. It has also been used in PBRs, where it has an active life of over 100 days, producing about two tonnes of 6-aminopenicillanic acid $kg^{-1}$of immobolised enzyme.

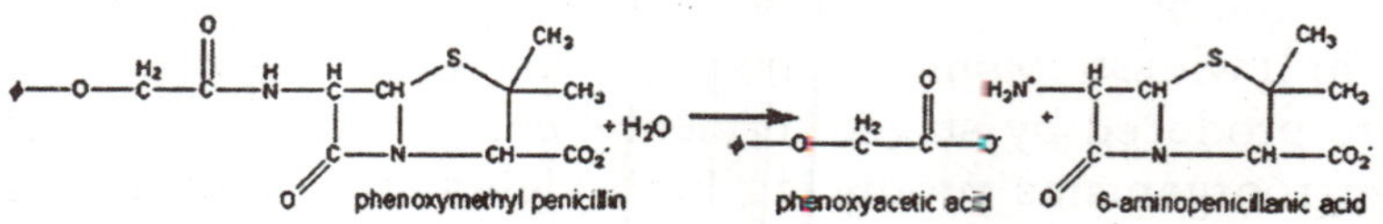

phenoxymethyl penicillin + H_2O → phenoxyacetic acid + 6-aminopenicillanic acid

The penicillin-G-amidases may be used 'in reverse' to synthesise penicillin and cephalosporin antibiotics by non - equilibrium kinetically controlled reactions. Ampicillin has been produced by the use of penicillin-G-amidase immobilised by adsorption to DEAE -cellulose in a packed bed column:

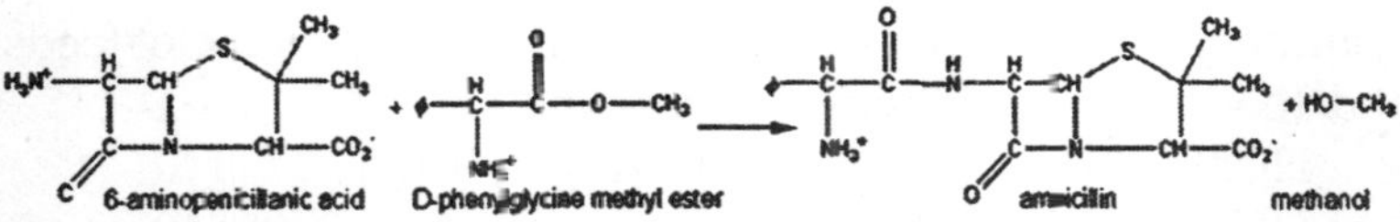

6-aminopenicillanic acid + D-phenylglycine methyl ester → ampicillin + $HO-CH_3$ methanol

Many other potential and proven antibiotics have been synthesised in this manner, using a variety of synthetic b-lactams and activated carboxylic acids.

PREPARATION OF ACRYLAMIDE

Acrylamide is an important monomer needed for the production of a range of economically useful polymeric materials. It may be produced by the addition of water to acrylonitrile.

$$CH_2{=}CHCN + H_2O \rightarrow CH_2{=}CHCONH_2$$

This process may be achieved by the use of a reduced copper catalyst (Cu^+); however, the yield is poor, unwanted polymerisation or conversion to acrylic acid ($CH_2{=}CHCOOH$) may occur at the relatively high temperatures involved (80 - 140°C) and the catalyst is difficult to regenerate. These problems may be overcome by the use of immobilised nitrile hydratase (often erroneously called a nitrilase) The enzyme from Rhodococcus has been used by the Nitto Chemical Industry Co. Ltd, as it contains only very low amidase activity which otherwise would produce unwanted acrylic acid from the acrylamide.

Immobilised nitrile hydratase is simply prepared by entrapping the intact cells in a cross-linked 10% (w/v) polyacrylamide/dimethylaminoethylmethacrylate gel and granulating the product.

It is used at 10°C and pH 8.0-8.5 in a semibatchwise process, keeping the substrate acrylonitrile concentration below 3% (w/v). Using 1% (w/v) immobilised-enzyme concentration (about 50,000 U l^{-1}) the process takes about a day.

Product concentrations of up to 20% (w/v) acrylamide have been achieved, containing negligible substrate and less than 0.02% (w/w) acrylic acid. Acrylamide production using this method is about 4000 tonnes per year. The closely related enzymes cyanidase and cyanide hydratase are used to remove cyanide from industrial waste and in the detoxification of feeds and foodstuffs containing amygdalin.

$HCN + 2H_2O \rightarrow HCOO^- + NH_4^+$

$HCN + H_2O \rightarrow HCONH_2$

IDENTIFYING USEFUL ANTIBIOTICS

Despite the wide variety of known antibiotics, less than 1% of antimicrobial agents have any medical or commercial value. The most commonly known antibiotic, Penicillin has a highly selective toxicity and therapeutic index (as eukaryotic animal cells do not contain peptidoglycan, they are usually unaffected by it). This is not so for many antibiotics.

Others simply lack advantage over the antibiotics already in use, or have no other practical applications.In order to identify the useful antibiotics, a process of screening is often employed.

Using this method, isolates of a large number of microorganisms are cultured and then tested for production of diffusible products which inhibit the growth of test organisms. However, the majority of the resulting antibiotics are already known and must therefore be disregarded.

The remainders must be tested for their selective toxicities and therapeutic activities and the best candidates can be examined and possibly modified.A more modern version of this approach is a rational design programme. This involves screening being directed towards finding new natural products that inhibit specific targets (e.g. a particular step of a metabolic pathway) on microorganisms, rather than tests to show general inhibition of a culture.

Industrial Production Techniques

Antibiotics are produced industrially by a process of fermentation, where the source microorganism is grown in large containers (100,000–150,000 liters or more) containing a liquid growth medium. Oxygen concentration, temperature, pH and nutrient levels must be optimal and are closely monitored and adjusted if necessary.

As antibiotics are secondary metabolites, the population

size must be controlled very carefully to ensure that maximum yield is obtained before the cells die.

Once the process is complete, the antibiotic must be extracted and purified to a crystalline product. This is simpler to achieve if the antibiotic is soluble in organic solvent. Otherwise it must first be removed by ion exchange, adsorption or chemical precipitation.

Strains used for Production

Microorganisms used in fermentation are rarely identical to the wild type. This is because species are often genetically modified to yield the maximum amounts of antibiotics. Mutation is often used and is encouraged by introducing mutagens such as ultraviolet radiation, x-rays or certain chemicals.

Selection and further reproduction of the higher yielding strains over many generations can raise yields by 20-fold or more. Another technique used to increase yields is gene amplification, where copies of genes coding for proteins involved in the antibiotic production can be inserted back into a cell, via vectors such as plasmids. This process must be closely linked with retesting of antibiotic production and effectiveness.

TYPES OF ANTIBIOTICS

The great number of diverse antibiotics currently available can be classified in different ways, e.g., by their chemical structure, their microbial origin, or their mode of action. They are also frequently designated by their effective range. Tetracyclines, the most widely used broad-spectrum antibiotics, are effective against both Gram-positive and Gram-negative bacteria, as well as against rickettsias and psittacosis-causing organisms.

Ciprofloxacin (Cipro) is another broad-spectrum antibiotic, effective in the treatment of mild infections of the urinary tract and sinuses.

The medium-spectrum antibiotics bacitracin, the erythromycins, penicillin and the cephalosporins are effective primarily against Gram-positive bacteria, although the streptomycin group is effective against some Gram-negative and Gram-positive bacteria. Polymixins are narrow-spectrum antibiotics effective against only a few species of bacteria.

Administration and Side Effects

Antibiotics are either injected, given orally, or applied to the skin in ointment form. Many, while potent anti-infective agents, also cause toxic side effects. Some, like penicillin, are highly allergenic and can cause skin rashes, shock and other manifestations of allergic sensitivity.

Others, such as the tetracyclines, cause major changes in the intestinal bacterial population and can result in superinfection by fungi and other microorganisms. Chloramphenicol, which is now restricted in use, produces severe blood diseases and use of streptomycin can result in ear and kidney damage. Many antibiotics are less effective than formerly because antibiotic-resistant strains of microorganisms have emerged.

Nonmedical Use

Antibiotics have found wide nonmedical use. Some are used in animal husbandry, along with vitamin B_{12}, to enhance the weight gain of livestock. However, some authorities believe the addition of antibiotics to animal feeds is `dangerous because continuous low exposure to the antibiotic can sensitize humans to the drug and make them unable to take the substance later for the treatment of infection.

In addition low levels of antibiotics in animal feed encourage the emergence of antibiotic-resistant strains of microorganisms. Drug resistance has been shown to be carried by a genetic particle transmissible from one strain of microorganism to another and the presence of low levels of antibiotics can actually cause an increase in the number of such particles in the bacterial population and increase the probability that such particles will be transferred to pathogenic, or disease-causing, strains. Antibiotics have also been used to treat plant diseases such as bacteria-caused infections in tomatoes, potatoes and fruit trees. The substances are also used in experimental research.

ENVIORNMENTAL BIOTECHNOLOGY

The Chambers Science and Technology Dictionary defines biotechnology as 'the use of organisms or their components in industrial or commercial processes, which can be aided by the techniques of genetic manipulation in developing e.g. novel plants for agriculture or industry.' Despite the inclusiveness of this definition, the biotechnology sector is still often seen as largely medical or pharmaceutical in nature, particularly amongst the general public.

While to some extent the huge research budgets of the drug companies and the widespread familiarity of their products makes this understandable, it does distort the full picture and somewhat unfairly so. However, while therapeutic instruments form, in many respects, the 'acceptable' face of biotechnology, elsewhere the science is all too frequently linked with unnatural interference.

While the agricultural, industrial and environmental applications of biotechnology are potentially very great, the shadow of Frankenstein has often been cast across them. Genetic engineering may be relatively commonplace in pharmaceutical thinking and yet in other spheres, like agriculture for example, society can so readily and thoroughly demonise it.

The history of human achievement has always been episodic. For a while, one particular field of endeavour seems to hold sway as the preserve of genius and development, before the focus shifts and development forges ahead in dizzy exponential rush in an entirely new direction.

So it was with art in the renaissance, music in the 18th century, engineering in the 19th and physics in the 20th. Now it is the age of the biological, possibly best viewed almost as a rebirth, after the great heyday of the Victorian naturalists, who provided so much input into the developing science.

It is then, perhaps, no surprise that the European Federation of Biotechnology begins its 'Brief History' of the science in the year 1859, with the publication of On the Origin of Species by Means of Natural Selection by Charles Darwin. Though his famous voyage aboard HMS Beagle, which led directly to the formulation of his (then) revolutionary ideas, took place when he was a young man, he had delayed making them known until 1858, when he made a joint presentation before the Linnaean Society with Alfred Russell Wallace, who had, himself, independently come to very similar conclusions.

Their contribution was to view evolution as the driving force of life, with successive selective pressures over time endowing living beings with optimised characteristics for survival. Neo-Darwinian thought sees the interplay of mutation and natural selection as fundamental.

The irony is that Darwin himself rejected mutation as too deleterious to be of value, seeing such organisms, in the language of the times, as 'sports' – oddities of no species benefit. Indeed, there is considerable evidence to suggest that he seems to have espoused a more Lamarckist view of biological progression, in which physical changes in an organism's lifetime were thought to shape future generations.

Darwin died in 1882. Ninety-nine years after his death, the first patent for a genetically modified organism was granted to Ananda Chakrabarty of the US General Electric, relating to a strain of Pseudomonas aeruginosa engineered to express the genes for certain enzymes in order to metabolise crude oil. Twenty years later still, in the year that saw the first working draft of the human genome sequence published and the announcement of the full genetic blueprint of the fruit fly, Drosophila melanogaster, that archetype of eukaryotic genetics research, biotechnology has become a major growth industry with increasing numbers of companies listed on the world's stock exchanges.

Thus, at the other end of the biotech timeline, a century and a half on from Origin of Species, the principles it first set out remain of direct relevance for what has been termed the 'chemical evolution' of biologically active substances and are commonly used in laboratories for in vitro production of desired qualities in biomolecules.

The Role of Environmental Biotechnology

While pharmaceutical biotechnology represents the glamorous end of the market, environmental applications are decidedly more in the Cinderella mould. The reasons for this are fairly obvious. The prospect of a cure for the many diseases and conditions currently promised by gene therapy and other biotech-oriented medical miracles can potentially touch us all. Our lives may, quite literally, be changed.

Environmental biotechnology, by contrast, deals with far less apparently dramatic topics and, though their importance, albeit different, may be every bit as great, their direct relevance is far less readily appreciated by the bulk of the population. Cleaning up contamination and dealing rationally with wastes is, of course, in everybody's best interests, but for most people, this is simply addressing a problem which they would rather had not existed in the first place.

Even for industry, though the benefits may be noticeable on the balance sheet, the likes of effluent treatment or pollution control are more of an inevitable obligation than a primary goal in themselves. In general, such activities are typically funded on a distinctly limited budget and have traditionally been viewed as a necessary inconvenience. This is in no way intended to be disparaging to industry; it simply represents commercial reality.

In many respects, there is a logical fit between this thinking and the aims of environmental biotechnology. For all the media circus surrounding the grand questions of our age, it is easy to forget that not all forms of biotechnology involve xenotransplantation, genetic modifica-tion, the use of stem cells or cloning.

Some of the potentially most beneficial uses of biological engineering and which may touch the lives of the majority of people, however indirectly, involve much simpler approaches. Less radical and showy, certainly, but powerful tools, just the same. Environmental biotechnology is fundamentally rooted in waste, in its various guises, typically being concerned with the remediation of contamination caused by previous use, the impact reduction of current activity or the control of pollution. Thus, the principal aims of this field are the manufacture of products in environmentally harmonious ways, which allow for the minimisation of harmful solids, liquids or gaseous outputs or the clean-up of the residual effects of earlier human occupation. The means by which this may be achieved are essentially two-fold.

Environmental biotechnologists may enhance or optimise conditions for existing biological systems to make their activities happen faster or more efficiently, or they resort to some form of alteration to bring about the desired outcome. The variety of organisms which may play a part in environmental applications of biotechnology is huge, ranging from microbes through to trees and all are utilised on one of the same three fundamental bases – accept, acclimatise or alter. For the vast majority of cases, it is the former approach, accepting and making use of existing species in their natural, unmodified form, which predominates.

The Scope for Use

There are three key points for environmental biotechnology interventions, namely in the manufacturing process, waste management or pollution control. Accordingly, the range of businesses to which environmental biotechnology has potential relevance is almost limitless. One area where this is most apparent is with regard to waste.

All commercial operations generate waste of one form or another and for many, a proportion of what is produced is biodegradable. With disposal costs rising steadily across the world, dealing with refuse constitutes an increasingly high contribution to overheads. Thus, there is a clear incentive for all businesses to identify potentially cost-cutting approaches to waste and employ them where possible.

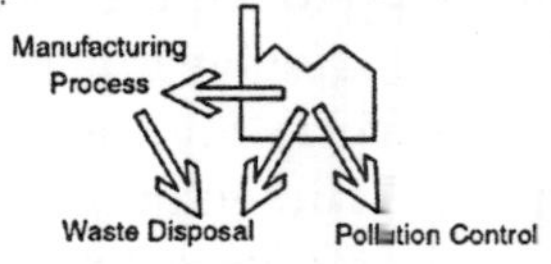

Fig. The Three Intervention Points

Changes in legislation throughout Europe, the US and elsewhere, have combined to drive these issues higher up the political agenda and biological methods of waste treatment have gained far greater acceptance as a result. For those industries with particularly high biowaste production, the various available treatment biotechnologies can offer considerable savings. Manufacturing industries can benefit from the applications of whole organisms or isolated biocomponents. Compared with conventional chemical processes, microbes and enzymes typically function at lower temperatures and pressures.

The lower energy demands this makes leads to reduced costs, but also has clear benefits in terms of both the environment and workplace safety. Additionally, biotechnology can be of further commercial significance by converting low-cost organic feedstocks into high value products or, since enzymatic reactions are more highly specific than their chemical counterparts, by deriving final substances of high relative purity.

Almost inevitably, manufacturing companies produce wastewaters or effluents, many of which contain biodegradable contaminants, in varying degrees. Though traditional permitted discharges to sewer or watercourses may be adequate for some, other industries, particularly those with recalcitrant or highly concentrated effluents, have found significant benefits to be gained from using biological treatment methods themselves on site.

Though careful monitoring and process control are essential, biotechnology stands as a particularly cost-effective means of reducing the pollution potential of wastewater, leading to enhanced public relations, compliance with environmental legislation and quantifiable cost-savings to the business.

Those involved in processing organic matter, for example, or with drying, printing, painting or coating processes, may give rise to the release of volatile organic compounds (VOCs) or odours, both of which represent environmental nuisances, though the former is more damaging than the latter. For many, it is not possible to avoid producing these emissions altogether, which leaves treating them to remove the offending contaminants the only practical solution. Especially for relatively low concentrations of readily water-soluble VOCs or odorous chemicals, biological technologies can offer an economic and effective alternative to conventional methods.

The use of biological cleaning agents is another area of potential benefit, especially where there is a need to remove oils and fats from process equipment, work surfaces or drains. Aside from typically reducing energy costs, this may also obviate the need for toxic or dangerous chemical agents. The pharmaceutical and brewing industries, for example, both have a long history of employing enzyme-based cleaners to remove organic residues from their process equipment.

In addition, the development of effective biosensors, powerful tools which rely on biochemical reactions to detect specific substances, has brought benefits to a wide range of sectors, including the manufacturing, engineering, chemical, water, food and beverage industries. With their ability to detect even small amounts of their particular target chemicals, quickly, easily and accurately, they have been enthusiastically adopted for a variety of process monitoring applications, particularly in respect of pollution assessment and control.

Contaminated land is a growing concern for the construction industry, as it seeks to balance the need for more houses and offices with wider social and environmental goals. The reuse of former industrial sites, many of which occupy prime locations, may typically have associated planning conditions attached which demand that the land be cleaned up as part of the development process.

With urban regeneration and the reclamation of 'brown-field' sites increasingly favoured in many countries over the use of virgin land, remediation has come to play a significant role and the industry has an ongoing interest in identifying cost-effective methods of achieving it. Historically, much of this has involved simply digging up the contaminated soil and removing it to landfill elsewhere.

Bioremediation technologies provide a competitive and

sustainable alternative and in many cases, the lower disturbance allows the overall scheme to make faster progress. As the previous brief examples show, the range of those which may benefit from the application of biotechnology is lengthy and includes the chemical, pharmaceutical, water, waste management and leisure industries, as well as manufacturing, the military, energy generation, agriculture and horticulture.

Clearly, then, this may have relevance to the viability of these ventures and, as was mentioned at the outset, biotechnology is an essentially commercial activity. Environmental biotechnology must compete in a world governed by the best practicable environmental option (BPEO) and the best available techniques not entailing excessive cost (Batneec).

Consequently, the economic aspect will always have a large influence on the uptake of all initiatives in environmental biotechnology and, most particularly, in the selection of methods to be used in any given situation. It is impossible to divorce this context from the decisionmaking process. By the same token, the sector itself has its own implications for the wider economy.

MARKET FOR ENVIRONMENTAL BIOTECHNOLOGY

The UK's Department of Trade and Industry estimated that 15–20% of the global environmental market in 2001 was biotech-based, which amounted to about 250–300 billion US dollars and the industry is projected to grow by as much as ten-fold over the following five years. This expected growth is due to greater acceptance of biotechnology for clean manufacturing applications and energy production, together with increased landfill charges and legislative changes in waste management which also alter the UK financial base favourably with respect to bioremediation.

Biotechnology-based methods are seen as essential to help meet European Union (EU) targets for biowaste diversion from landfill and reductions in pollutants. Across the world the existing regulations on environmental pollution are predicted to be more rigorously enforced, with more stringent compliance standards implemented. All of this is expected to stimulate the sales of biotechnology-based environmental processing methods significantly and, in particular, the global market share is projected to grow faster than the general biotech sector trend, in part due to the anticipated large-scale EU aid for environmental clean-up in the new accession countries of Eastern Europe.

Other sources paint a broadly similar picture. The BioIndustry Association (BIA) survey, Industrial Markets for UK Biotechnology – Trends and Issues, published in 1999 does not quote any monetary sector values per year, but gives the size of the UK sector as employing 40 000 people in 1998 with an average yearly growth over 1995–98 of 20%.

Environmental biotech is reported as representing around 10% of this sector. An Arthur Anderson report of 1997 gives the turnover of the UK biotech sector as 702 million pounds sterling in 1995/96, with a 50% growth over three years. A 1998 Ernst and Young report on the European Life Sciences Sector says that the market for biotechnology products has the potential to reach 100 billion pounds sterling worldwide by 2005.

The Organisation for Economic Cooperation and Development (OECD) estimates that the global market for environmental biotechnology products and services alone will rise to some US$75 billion by the year 2000, accounting for some 15 to 25% of the overall environmental technology market, which has a growth rate estimated at 5.5% per annum. The UK potential market for environmental biotechnology products and services is estimated at between 1.65 and 2.75 billion US dollars and the growth of the sector stands at 25% per annum, according to the Bio- Commerce Data European Biotechnology Handbook.

The world market size of biotechnology products and services was estimated to be approximately 390 billion US dollars in the year 2000. The benefits are not, however, confined to the balance sheet. The Organisation for Economic Cooperation and Development (OECD 2001) concluded that the industrial use of biotechnology commonly leads to increasingly environmentally harmonious processes and additionally results in lowered operating and/or capital costs. For years, industry has appeared locked into a seemingly unbreakable cycle of growth achieved at the cost of environmental damage.

The OECD investigation provides what is probably the first hard evidence to support the reality of biotechnology's long-heralded promise of alternative production methods, which are ecologically sound and economically efficient. A variety of industrial sectors including pharmaceuticals, chemicals, textiles, food and energy were examined, with a particular emphasis on biomass renewable resources, enzymes and bio-catalysis.

While such approaches may have to be used in tandem with other processes for maximum effectiveness, it seems that their use invariably leads to reduction in operating or capital costs, or both. Moreover, the research also concludes that it is clearly in the interests of governments of the developed and developing worlds alike to promote the use of biotechnology for the substantial reductions in resource and energy consumption, emissions, pollution and waste production it offers. The potential contribution to be made by the appropriate use of biotechnology to environmental and economic sustainability would seem to be clear.

The upshot of this is that few biotech companies in the environmental sector perceive problems for their own business development models, principally as a result of the wide range of businesses for which their services are applicable, the relatively low market penetration to date and the large potential for growth.

Competition within the sector is not seen as a major issue either, since the field is still largely open and unsaturated. Moreover, there has been a discernible tendency in recent years towards niche specificity, with companies operating in more specialised subarenas within the environmental biotechnology umbrella. Given the number and diversity of such possible slots, coupled with the fact that new opportunities and the technologies to capitalise on them, are developing apace, this trend seems likely to continue. It is not without some irony that companies basing their commercial activities on biological organisms should themselves come to behave in such a Darwinian fashion. However, the picture is not entirely rosy.

Typically the sector comprises a number of relatively small, specialist companies and the market is, as a consequence, inevitably fragmented. Often the complexities of individual projects make the application of 'standard' off-theshelf approaches very difficult, the upshot being that much of what is done must be significantly customised. While this, of course, is a strength and of great potential environmental benefit, it also has hard commercial implications which must be taken into account.

A sizeable proportion of companies active in this sphere, have no products or services which might reasonably be termed suitable for generalised use, though they may have enough expertise, experience or sufficiently perfected techniques to deal with a large number of possible scenarios. The fact remains that one of the major barriers to the wider uptake of biological approaches is the high perceived cost of these applications.

Part of the reason for this lies in historical experience. For many years, the solutions to all environmental problems were seen as expensive and for many particularly those unfamiliar with the multiplicity of varied technologies available, this has remained the prevalent view. Generally, there is often a lack of financial resource allocation available for this kind of work and biotech providers have sometimes come under pressure to reduce the prices for their services as a result.

Greater awareness of the benefits of biotechnology, both as a means to boost existing markets and for the opening up of new ones, is an important area to be addressed. Many providers, particularly in the UK, have cited a lack of marketing expertise as one of the principal barriers to their exploitation of novel opportunities.

In addition, a lack of technical understanding of biotech approaches amongst target industries and, in some cases, downright scepticism regarding their efficacy, can also prove problematic. Good education, in the widest sense, of customers and potential users of biological solutions will be one major factor in any future upswing in the acceptance and utilisation of these technologies.

Modalities and Local Influences

Another of the key factors affecting the practical uptake of environmental biotechnology is the effect of local circumstances. Contextual sensitivity is almost certainly the single most important factor in technology selection and represents a major influence on the likely penetration of biotech processes into the marketplace.

Neither the nature of the biological system, nor of the application method itself play anything like so relevant a role. This may seem somewhat unexpected at first sight, but the reasons for it are obvious on further inspection. While the character of both the specific organisms and the engineering remain essentially the same irrespective of location, external modalities of economics, legislation and custom vary on exactly this basis.

Accordingly, what may make abundant sense as a biotech intervention in one region or country, may be totally unsuited to use in another. In as much as it is impossible to discount the wider global economic aspects, disassociating political, fiscal and social conditions equally cannot be done, as the following example illustrates.

In 1994, the expense of bioremediating contaminated soil in the United Kingdom greatly exceeded the cost of removing it to landfill. Six years later, with successive changes of legislation and the imposition of a landfill tax, the situation has almost completely reversed. In those other countries where landfill has always been an expensive option, remediation has been embraced far more readily.

While environmental biotechnology must, inevitably, be viewed as contextually dependent, as the previous example shows, contexts can change. In the final analysis, it is often fiscal instruments, rather than the technologies, which provide the driving force and sometimes seemingly minor modifications in apparently unrelated sectors can have major ramifications for the application of biotechnology.

The legal framework is another aspect of undeniable importance in this respect. Increasingly tough environmental law makes a significant contribution to the sector and changes in regulatory legislation are often enormously influential in boosting existing markets or creating new ones.

When legislation and economic pressure combine, as, for example, they have begun to do in the European Landfill Directive, the impetus towards a fundamental paradigm shift becomes overwhelming and the implications for relevant biological applications can be immense. There is a natural tendency to delineate, seeking to characterise technologies into particular categories or divisions. However, the essence of environmental biotechnology is such that there are many more similarities than differences.

Though it is, of course, often helpful to view individual technology uses as distinct, particularly when considering treatment options for a given environmental problem, there are inevitably recurrent themes which feature throughout the whole topic.

Moreover, this is a truly applied science. While the importance of the laboratory bench cannot be denied, the controlled world of research translates imperfectly into the harsh realities of commercial implementation. Thus, there can often be a dichotomy between theory and application and it is precisely this fertile ground which is explored in the present work.

In addition, the principal underlying approach of specifically environmental biotechnology, as distinct from other kinds, is the reliance on existing natural cycles, often directly and in an entirely unmodified form. Thus, this science stands on a foundation of fundamental biology and biochemistry. To understand the application, the biotechnologist must simply examine the essential elements of life, living systems and ecological circulation sequences.

However engineered the approach, this fact remains true.

In many respects, environmental biotechnology stands as the purest example of the newly emergent bioindustry, since it is the least refined, at least in terms of the basis of its action.

In essence, all of its applications simply encourage the natural propensity of the organisms involved, while seeking to enhance or accelerate their action. Hence, optimisation, rather than modification, is the typical route by which the particular desired end result, whatever it may be, is achieved and, consequently, a number of issues feature as common threads of individual technologies.

Integrated Approach

ntegration is an important aspect for environmental biotechnology. One theme that will be developed throughout this book is the potential for different biological approaches to be combined within treatment trains, thereby producing an overall effect which would be impossible for any single technology alone to achieve.

However, the wider goal of integration is not, of necessity, confined solely to the specific methods used. It applies equally to the underpinning knowledge that enables them to function in the first place and an understanding of this is central to the rationale behind this book. In some spheres, traditional biology has become rather unfashionable and the emphasis has shifted to more exciting sounding aspects of life science.

While the new-found concentration on 'Ecological processes', or whatever, sounds distinctly more 'Environmental', in many ways and somewhat paradoxically, it sometimes serves the needs of environmental biotechnology rather less well. The fundamentals of living systems are the stuff of this branch of science and, complex though the whole picture may be, at its simplest the environmental biotechnologist is principally concerned with a relatively small number of basic cycles.

In this respect, a good working knowledge of biological processes like respiration, fermentation and photosynthesis, a grasp of the major cycles by which carbon, nitrogen and water are recycled and an appreciation of the flow of energy through the biosphere must be viewed as prerequisites. The intent here has been neither to insult the readership by parading what is already well known, nor gloss over aspects which, if left unexplained, at least in reasonable detail, might only serve to confuse.

However, this is expressly not designed to be a substitute for much more specific texts on these subjects, nor an entire alternative to a cohesive course on biology or biochemistry. The intention is to introduce and explain the necessary aspects and elements of various metabolic pathways, reactions and abilities as required to advance the reader's understanding of this particular branch of biotechnology.

A large part of the reasons for approaching the subject in this way is the fact that there really is no such thing as a 'typical' environmental biotechnologist. Practitioners come into the profession from a wide variety of disciplines and by many different routes. Thus, amongst their ranks are agronomists, biochemists, biologists, botanists, enzymologists, geneticists, microbiologists, molecular biologists, process engineers and protein technologists, all of whom bring their own particular skills, knowledge base and experiences.

The applied nature of environmental biotechnology is obvious. While the science underlying the processes themselves may be as pure as any other, what distinguishes this branch of biological technology are the distinctly real-life purposes to which it is put. Hence, part of the intended function of this book is to attempt to elucidate the former in order to establish the basis of the latter.

At the same time, as any applied scientist will confirm, what happens in the field under operational conditions represents a distinct compromise between the theoretical and the practically achievable. At times, anything more than an approximation to the expected results may be counted as something of a triumph of environmental engineering.

Closing Remarks

The celebrated astronomer and biologist, Sir Fred Hoyle,

said that the solutions to major unresolved problems should be sought by the exploration of radical hypotheses, while simultaneously adhering to well-tried and tested scientific tools and methods. This approach is particularly valid for environmental biotechnology. With new developments in treatment technologies appearing all the time, the list of what can be processed or remediated by biological means is ever changing.

By the same token, the applications for which biotechnological solutions are sought are also subject to alteration. For the biotech sector to keep abreast of these new demands it may be necessary to examine some truly 'radical hypotheses' and possibly make use of organisms or their derivatives in ways previously unimagined. This is the basis of innovation; the inventiveness of an industry is often a good measure of its adaptability and commercial robustness.

MICROBIAL BIOMASS

THE AQUATIC NITROGEN CYCLE

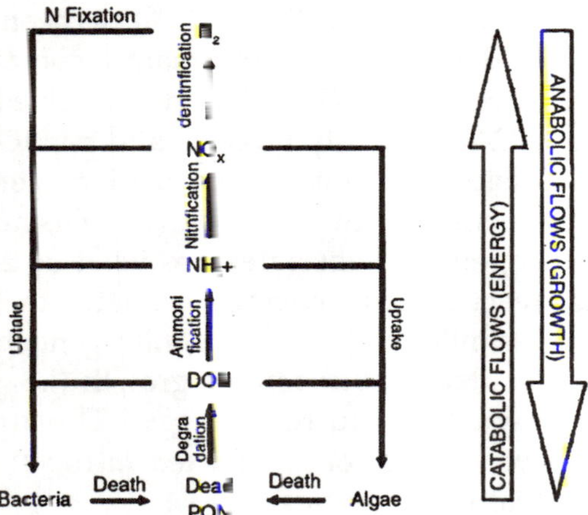

Fig. General Overview of Nitrogen Transformation Processes within the Nitrogen Cycle, Including Nitrogen flows Through Bacteria and algae

The aquatic nitrogen cycle is a complex system consisting

of several different forms of nitrogen with different oxidation states that are interconnected by numerous conversion reactions.

The different components of the aquatic nitrogen cycle can be separated from each other in space and/or time. The nitrogen cycle plays an important role in aquatic systems since nitrogen is usually the element limiting microbial production in these systems. However, due to its complexity, various aspects of the nitrogen cycle are still poorly understood.

Bioavailable nitrogen is present in the environment as "free" nitrogen (atmospheric N_2) and "fixed" nitrogen. The latter can roughly be divided into particulate nitrogen (functionally defined as all nitrogenous material that can be captured by filtration) and dissolved nitrogen (all nitrogenous material that passes through a filter, typically GF/F).

The dissolved fixed nitrogen pool can be subdivided into dissolved inorganic nitrogen (DIN) and dissolved organic nitrogen (DON). The DIN pool includes ammonium (NH^{4+}), nitrate (NO^{3-}) and nitrite (NO^{2-}) of which the last two can be taken together as NOx-. The DON pool is a more complex pool of which only a small fraction (< 20%) has been chemically identified. The identifiable fraction mainly consists of urea, dissolved free amino acids (DFAAs) and dissolved combined amino acids (DCAAs, mainly proteins and peptides).

The conversion reactions within the nitrogen cycle can have two "Directions". In catabolic (or "Dissimila-tory") reactions, nitrogenous substrates are used as a source of energy, these reactions are generally mediated by bacteria. In anabolic (or "assimilatory") reactions, nitrogenous substrates are used as a source of nitrogen for growth (i.e. production of biomass), which requires energy. The uptake and incorporation of various forms of fixed nitrogen by natural aquatic microbial communities.

Microbial Nitrogen Uptake

Natural aquatic microbial communities (mainly bacteria and algae) can use various forms of nitrogen as nitrogen sources for growth. Although N_2 is abundantly present in the atmosphere, its importance as a nitrogen source for aquatic microbial communities is limited, especially in systems with high availability of fixed nitrogen sources, because the use of nitrogen from N_2 for growth ("Nitrogen fixation") requires a high energy investment and only a limited group of organisms (cyanobacteria and some heterotrophic bacteria) is capable of nitrogen fixation Therefore, the main nitrogen sources for microbial communities in coastal waters and sediments are DIN and DON.

Traditionally, studies on aquatic microbial nitrogen uptake focused on uptake of DIN by algae while bacteria were only considered as remineralizers of organic matter. However, the present view of microbial nitrogen uptake is considerably more complex since DON (especially amino acids and urea) has been found to be a potentially important nitrogen source as well and because it is now clear that bacteria are also important players in aquatic microbial nitrogen uptake. In addition to bacteria and algae, aquatic microbial communities can also include Archaea. However, at present their role in total microbial nitrogen uptake is unknown.

The pathways for uptake of dissolved nitrogen by algae and bacteria are schematized. The uptake process consists of two steps that are not necessarily coupled. The first step is the uptake of dissolved nitrogenous compounds from the surrounding water into the cells. Small molecules like NH4+, NOx- and urea can easily be taken up into cells while larger compounds like proteins and peptides first need to be broken down into individual amino acids that can either be directly taken up into the cells or be broken down to NH4+ which is then taken up into the cells.

The second step is incorporation of the nitrogen in new biomass. For this, intracellular NO^{3-} and urea are first converted to NH^{4+} which is used to synthesize amino acids that form the building blocks for proteinaceous biomass components. Since amino acids can directly be incorporated into biomass after uptake (i.e. little or no conversion is required), these are a highly preferred nitrogen source. Conversely, the microbial preference for NO^{x-} as a nitrogen source is low because of the relatively high energy investment required for its conversion to NH^{4+}. Preferences for NH4+ and urea are generally in between those for amino acids and NO^{x-}.

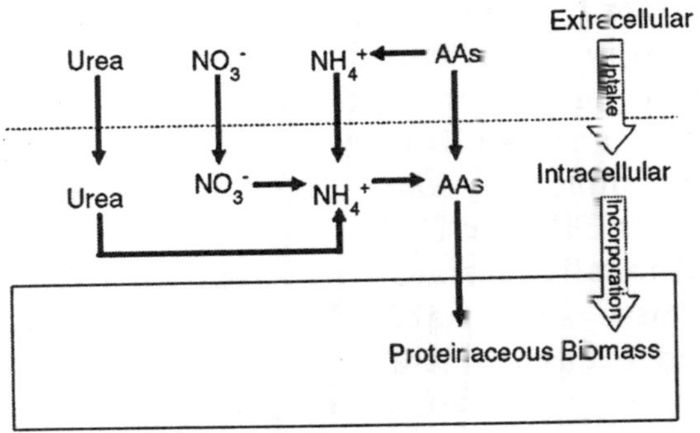

Fate of Microbial Nitrogen

An interesting, yet poorly studied, aspect of the aquatic nitrogen cycle is the fate of microbial nitrogen (i.e., what happens with the nitrogen after it has been incorporated into microbial biomass?) which is coupled with the fate of other elements in microbial biomass such as carbon. Although nitrogen can be retained in microbial biomass during continued growth (by cell division), microbial biomass (especially that of bacteria) is often relatively constant over time, meaning that production of new biomass is balanced by loss of biomass due to cell death.

Death of microorganisms can be due to various causes such as viral lysis, predation and deteriorating environmental conditions. After cell death, the leftovers (remnants) of microbial biomass are subject to degradation. During degradation, nitrogen is released back into the system as DON and/or DIN that can be used as nitrogen source for production of new microbial biomass or be lost from the system.

Some microbial biomass components are more resistant to degradation ("Refractory") than others and may accumulate in the system where they serve as a long term sink for microbial-derived nitrogen. Another potential fate of microbial nitrogen is transfer to higher trophic levels of the food web when microbial biomass is consumed by larger organisms.

Stable Isotopes in Ecological and Biogeochemical

Elements such as carbon (C) and nitrogen (N) can occur in different forms called isotopes. Different isotopes of a single element have the same number of protons but a different number of neutrons. Some isotopes are radioisotopes, meaning that these isotopes are subject to radioactive decay. Radioisotopes have been used in ecological and biogeochemical studies (for example for measuring primary production using ^{14}C).

However, work with radioisotopes comes with a potential health risk and can be restricted by law. Moreover, there is no convenient radioisotope of nitrogen which makes that the use of radioisotopes in studies on nitrogen cycling is restricted to work with the short-lived radioisotope ^{13}N.

Isotopes that are not subject to radioactive decay are stable isotopes. A single element can have two or more stable isotopes, of which one is by far the most abundant (e.g. ^{12}C and ^{14}N) while the other forms are rare (for example: 99.6% of all nitrogen in nature is 14N and only 0.4% is ^{15}N).

The abundance of these rare isotopes is measured by isotope ratio mass spectrometry (IRMS), a technique that uses a magnetic field to separate gas molecules (CO_2 in case of ^{13}C analysis and N_2 in case of ^{15}N analysis) of different molecular weights resulting from different stable isotope compositions. The relative abundance of these molecules with different molecular weights can be used to determine the relative abundance of the rare (or "Heavy") isotope.

This abundance is typically expressed by the delta notation (δ) which is the ratio between the rare ("Heavy") isotope and the common (or "Light") isotope relative to that of a standard in permille (%°) units.

Stable isotopes can be used in ecological and biogeochemical studies in two different ways. The first way is to study the natural abundance of the rare, heavy isotope. Here, the relevant information lies in small differences in relative abundance of the heavy isotope between different compartments of a system.

Table. Carbon and Nitrogen Isotopes

		Type of Isotope	*Natural Abundance*
Carbon	10_C	Radio (½ life 17 Sec)	
	12_C	Stable	98.9%
	13_C	Stable	1.1%
	13_C	Radio (½ life 5730 yr)	
Nitrogen	13_N	Radio (½ life 10 min)	
	14_N	Stable	99.6%
	15_N	Stable	0.4%
	16_N	Radio (½ life 7 Sec)	

These differences are the results of isotopic fractionation, which involves changes in the natural abundance of the heavy isotope during reactions caused by small differences in mass-dependent reaction rates of heavy and light isotopes. The second way to use stable isotopes is by stable isotope labeling.

In this approach, a substrate that is highly enriched in the heavy isotope is deliberately added to a system after which the abundance of the heavy isotope (the "Label") is measured in various components of the system which allows to trace and quantify flows of the label trough these components.

The material presented in this thesis all concerns stable isotope labeling work with the nitrogen isotope 15N and the carbon isotope 13C.

BIOMARKERS

Biomarkers are compounds that are unique to a specific group of organisms. Ideally, biomarker compounds are part of the organism's biomass and make up a relatively constant fraction of that biomass so that biomarker concentrations can be converted to total biomass of the group it represents. A variety of compounds have been used as biomarkers for various microbial groups.

Some of these compounds are rapidly degraded after death of the organism, which makes them specific for living biomass ("ecological biomarkers"), for example phospholipid-derived fatty acids (PLFAs). Other components are more resistant to degradation ("biogeochemical biomarkers"), for example D-amino acids from bacterial cell walls, which makes that these components can accumulate in a system, particularly in sediment where the abundance of these biomarkers can provide information on the contribution of material from the corresponding group of organisms to the total organic matter pool in the sediment. A third group of biomarkers are highly refractory molecules ("geological biomarkers").

The concentrations and characteristics of these "molecular fossils" in sediments can be used for paleoenvironmental reconstructions. A valuable application of biomarkers in ecological- and biogeochemical studies is analysis of their stable isotope composition, which requires compound-specific stable isotope analysis.

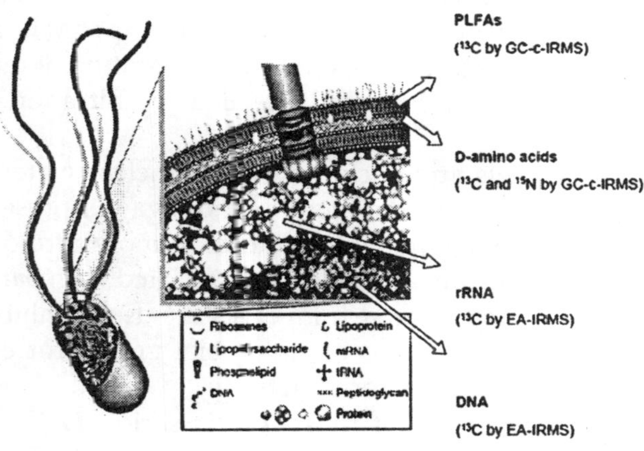

Fig. A Selection of Bacterial Biomarkers used in Combination with Stable Isotope Analysis.

This can be done at natural abundance level or in combination with stable isotope labeling where the latter provides the unique opportunity to directly link microbial identity (biomarker), biomass (biomarker concentration) and activity (label incorporation).

Selection of bacterial biomarkers that have been used in combination with stable isotope analysis. Biomarkers and stable isotope analysis of biomarkers, including some examples of applications in microbial ecology, can be found in Boschker and Middelburg while technical aspects of compound-specific stable isotope analysis.

The Problem

Microbial uptake of nitrogen from various nitrogenous substrates is generally studied by analysis of uptake of 15N from 15N-labeled substrates. This approach has frequently been used in water column studies where 15N enrichment of total particulate material in the water is considered to represent total microbial 15N uptake.

However, more detailed assessment of 15N uptake by specific microbial groups (bacteria versus algae) has long been hampered by methodological problems. Until recently, two different approaches have been used to discriminate 15N uptake by bacteria versus algae. The first approach is size fractionation, which concerns physical separation of relatively small bacteria from relatively large algae by filtration or flow cytometry.

Although this approach may work in clear waters when a size difference between bacteria and algae is present, size fractionation is not suitable for studies in more turbid waters where bacteria and algae are closely associated with particulate matter. The second approach involves the use of inhibitors to eliminate activity of a specific microbial group, for example antibiotics to inhibit bacterial activity.

However, the use of inhibitors is seriously limited by uncertainties concerning both the specificity and efficiency of inhibitors. Next to the problems in water column studies presented above, studies on microbial nitrogen uptake in sediments have been hampered by even more fundamental methodological problems because of the inability to distinguish between ^{15}N taken up by the microbial community and the remaining added ^{15}N-substrate and/or possible other ^{15}N-labeled extracellular compounds bound to the sediment.

Moreover, another limitation of traditional ^{15}N-labeling work is that it does not allow clean assessment of immobilization of dissolved nitrogen into microbial biomass (the relevant process from a nitrogen cycling perspective as well as from a food web point of view) since total ^{15}N "Uptake" also includes uptake of nitrogenous substrates into microbial cells, which is not necessarily directly coupled to incorporation into biomass. The method presented in this thesis is a new approach that has the potential to overcome the problems mentioned above and therefore may help to further clarify nitrogen flows through bacteria and algae in natural microbial communities.

The Solution

The method presented in this thesis involves the use of compound-specific stable isotope analysis of biomarkers. This approach has already been used at NIOO-CEME for some years in ^{13}C-labeling studies where analysis of ^{13}C incorporation into various group-specific PLFAs provided valuable information on ^{13}C flows through different microbial groups, including bacteria and algae. However, since PLFAs do not contain nitrogen, another type of compounds was required for 15Nlabeling studies.

This thesis investigates the potential of 15N analysis of hydrolysable amino acids (HAAs), including the bacterial biomarker D-alanine (D-Ala). HAAs are amino acids that can be released from proteinaceous biomass (proteins and protein-like material) by hydrolysis in hot acid. HAAs make up a large fraction (50- 80%) of total microbial biomass meaning that analysis of 15N incorporation into total HAAs (THAAs) should provide a good indication of 15N incorporation into total microbial biomass. Amino acids can occur as D- and L-stereoisomers. Most HAAs are L-amino acids that are common biomass constituents of all organisms and therefore have very limited biomarker potential

Conversely, hydrolysable D-amino acids are rare and only occur in bacterial biomass, more specifically in peptidoglycan, a structural component of bacterial cell walls. Therefore, these D-amino acids can be used as bacterial biomarkers. Although different D-amino acids can be present in peptidoglycan, our method focuses on D-Ala because this is the only D-amino acid that is present in all bacteria and because it makes up a relatively stable fraction of total bacterial biomass. Moreover, D-Ala has been shown to be most suitable for stable isotope analysis by GC-c-IRMS.

The method can also be used to measure incorporation of the carbon isotope 13C into D-Ala and other HAAs, which allows assessment of carbon flows through bacteria and algae. This is particularly interesting in direct combination with analysis of nitrogen flows.

Chapter 10

Tissue Culture and Laser Application

To study the effects of conventional laser application on the retinal pigment epithelium (RPE) in a perfusion tissue culture model of porcine retinal pigment epithelium without overlying neurosensory retina.

RPE with underlying choroid was prepared from enucleated porcine eyes and fixed in a holding ring. Specimens were then placed in two-compartment tissue culture containers and were cultured during continuous perfusion with culture medium at both sides of the entire specimen, the upper RPE and the lower choroid (12 specimens out of 6 eyes).

Cultures were kept for 1, 3, 7 and 14 days and were examined histologically. Laser treatment was performed on each tissue ring by application of 3 ! 3 laser burns one day after culture began (argon ion laser, wavelength: 514 nm, pulse duration: 100 ms; spot size: 200 Ìm) using different energy levels (400–1,000 mW); (16 specimens out of 8 eyes).

During laser treatment a marked lightening of the RPE with centrifugal spreading was observed. Using higher levels of energy, a contraction of the RPE towards the centre of the laser spot was noticed. One day after laser photocoagulation histology revealed destruction of RPE; within 3–7 days of culture, migration and proliferation of neighboring cells was observed in several lesions.

After 7 days the initial defect of the irradiated area was covered with dome shaped RPE cells and after 14 days multilayered RPE cells were showing ongoing proliferation.

However, there were also cases without proliferation after laser treatment.

The non-treated, continuously perfused RPE showed regular appearance in histological sections: during the first 7 days of culture, light microscopy revealed a normal matrix with a well-differentiated RPE monolayer. Subsequently proliferation even without treatment was observed and after 14 days the RPE became multilayered.

It was possible to study the early healing response to the effect of laser treatment using the permanently perfused tissue culture system. A marked proliferation and repair of the laser defect could be observed in several but not all lesions. After 14 days even without laser treatment a proliferative multilayered RPE was present. Although this limits the use of the system for longer than 7 days, it seems to be useful for investigation of RPE-related disorders.

The retinal pigment epithelium (RPE) is a complex cell layer involved in many eye diseases such as age-related macular disease (AMD), diabetic maculopathy (DMP), proliferative diabetic retinopathy (DRP), or central serous retinopathy (CSR). In these diseases laser photocoagulation is known to have a therapeutic effect and is therefore widely used.

Despite this widespread clinical use, the exact mechanism of laser photocoagulation is not completely understood. It is known that the therapeutic effect derives not from the laser burn itself, but from the subsequent biological reaction. Laser application using 514 nm is absorbed up to 50 per cent by RPE cells resulting in damages with subsequent healing and therapeutic effects, widely used clinically today.

The ability to maintain RPE cells in culture has important advantages for the understanding of the pathobiology of several ocular diseases. Especially in organ cultures, the interaction of different tissue cells can be studied. The main difficulty encountered when culturing RPE is to retain an intact monolayer over a long period of time, which would be helpful for elucidating complete pathomechanisms of several diseases. However, most previous studies are of limited value because the RPE cells were maintained on artificial substratum (e.g. polystyrene culture disks) although it has been well established that the substratum has pronounced effects on cell behaviour. To avoid these limitations

Del Priore and coworkers first developed an organ culture system to maintain monolayers of RPE still attached to Bruch's membrane which provided the opportunity to study the response of the RPE to external stimuli such as photocoagulation.

Using this model it was demonstrated that laser photocoagulation led to an acute disruption of individual RPE cells and a separation of damaged RPE cells from Bruch's membrane. Treated areas were covered with irregular mounds of RPE cells within 7 days, which was supposed to mimic the response of RPE following laser photocoagulation in vivo.

For our approach to study laser tissue interaction a new model for cultivating RPE cells was employed. It was developed to cultivate tissue in an organotypical environment of a two-compartment system under permanent perfusion with medium. Good results for preservation of adult full-thickness retinal tissue using this model were presented recently. In a second step we used this model to maintain RPE as an organ culture over a longer period of time and to characterize the vitality of this system by irradiating the RPE monolayer with a conventional argon laser and performing light microscopy afterwards.

TISSUE PREPARATION

Fresh enucleated porcine eyes were prepared with the use of an operation-microscope under sterile conditions according to the method developed by Kobuch et al. using the MinuCell-perfusion system.

Thus 14 eyes were cut parallel to the limbus at a distance of 4 mm behind the iris-lens-diaphragm and the whole anterior cap was removed and most of the adherent vitreous. Next the posterior eye cup was separated into two equal parts by dissecting along the median raphe and the optic disc with a sharp knife.

Each part was shortened at the edges to become a flat specimen of sclera, choroid, RPE and retina with partly adherent vitreous. The specimen was fixed to a flat surface with pins.

Neurosensory retina was peeled off from the RPE with tweezers. Small scissors were used to separate the choroid with the overlying structures from the sclera leaving a layer of choroid and RPE. In order to cultivate this matrix we used a ring holder system, placed into a perfusion container which has been previously described in detail. Briefly, the cell carrier consisted of a black holding ring, in which the cell layer was placed.

The internal diameter was 9 mm and the external diameter 13 mm. To fix the support, a white span ring was pressed into the holding ring.

Thus the white ring was moved under the choroid and the black ring was put onto the RPE in order to fix the specimen into this ring system thus having choroid at the lower and RPE at the upper side. This was then placed into the container which was designed for one to six ring systems.

By placing the ring system into one of the compartments of the container, two compartments were created separated by the tissue and the ring holder system. Both compartments had two openings to be perfused with medium. The use of the gradient container allows permanent perfusion of different media at the upper (retina) and lower (choroid) side of the sheets. In our study the same medium was used at both sides.

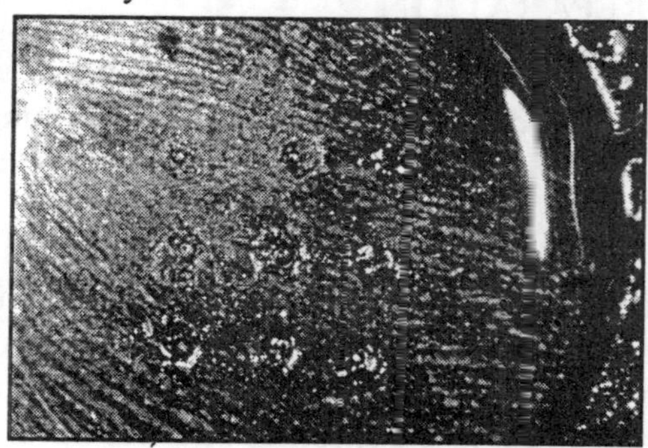

Fig. Tissue Holder Ring System with Coagulated RPE-Monolayer on the Top. Nine Laser Lesions can be Noticed

Over silicone tubes and luer fittings the container was

connected to specific screw caps and media bottles allowing the permanent perfusion of the tissue.

The system ran with a peristaltic pump with a speed of 1 ml/h on a warming table (40°C) out of the atmosphere of a CO2-incubator. DMEM (Life Technologies) with 25 mmol HEPES, 15 per cent porcine serum and 1 per cent penicillin- streptomycin out of 10.000 IU was used.

The specimens were cultivated under these conditions for up to 1, 3, 7 and 14 days. Additionally one specimen was cultured up to 28 days. All were examined histologically by light microscopy.

Table. Number of examined specimens during the study differentiated by perfused RPE without laser treatment and perfused RPE with laser treatment: every specimen treated by laser photocoagulation covered nine spots

	1 day	3 days	7 days	14 days
Perfused RPE without laser	3	3	3	3
Perfused RPE with laser	4	4	4	4

LASER APPLICATION

Using a special single chamber it was possible to perform laser photocoagulation to the RPE by positioning the chamber perpendicularly to the laser beam. Laser treatment spots (distance 200m) per tissue ring system was performed one day after cultivation using a conventional argon laser [wavelength: 514 nm, pulse duration: 100 ms, spot size: 200 Ìm (even on tissue), power: 400–1,000 mW]. Different energy levels were utilized according to different pigmentation of the tissue.

Thus energy was chosen in a way that a laser effect could be recognized ophthalmoscopically as a gray lesion. Also noticed was some pigment which was released from the RPE during the laser application and always a centrifugal enlarging lightening of the RPE could be observed in the first milliseconds after irradiation.

The lesion appearance varied according to the laser power, however, the overall variability of the ophthalmoscopic aspect of the lesions was considered to be minor. The irradiated tissue was kept under perfusion conditions for 1, 3, 7 and 14 days and examined histologically afterwards by light microscopy.

STUDY DESIGN

The specimens were divided into two subgroups: perfused RPE without laser treatment and perfused RPE with laser treatment. Each laser-treated specimen received nine laser spots having a distance to each other of about 200 Ìm.

LIGHT MICROSCOPY

Tissue fixation was performed in formalin 10 per cent for paraffin embedding to get a first impression and then in 3 per cent glutaraldehyde (Sigma) in DMEM without porcine serum. Fixation took place after about 24 h. Specimens then were embedded with EPON.

Perfused RPE without Laser Application

Under perfusion untreated specimens were well differentiated. Regular RPE formation with apical melanin was seen 1 day and 3 days after cultivation. The RPE cells were in close apposition and some microvilli at the surface of the RPE layer were recognized. After 7 days the RPE formation changed.

Apical melanin was still present but marked alterations of the RPE cells occurred yielding a cuboidal and partly dome shaped morphology. Some vacuoles could be found inside the cell suggesting phagocytosis.

After 14 days of perfusion culture the RPE was multilayered. After 28 days further growth of the multilayered RPE was noticed in one specimen. In all histologic sections the choroid showed a good preservation of the matrix with well differentiated vessels having an intact endothelium.

Perfused RPE with Laser Application

It was difficult to identify suspected areas as proper laser lesions. In all specimens examined only a minority of the laser spots applied were found by sectioning the tissue. Thus many areas which were suspected to be laser lesions did not show cell proliferation a few days after irradiation; however, in these cases differentiation from artifacts by preparation was difficult. Thus for all given times of examination just six laser lesions could be clearly identified as areas treated by laser having following appearances: One day after laser treatment the irradiated area could be identified as an area devoid of RPE cells.

Bruch's membrane remained intact judged by light microscopy but appeared wrinkled over the whole length of the laser burn. The underlying choroid had closed vessels, extra cellular pigment and cell debris. After 3 days, migration and proliferation of cells derived from the RPE next to the lesion could be noticed.

These cells were flat and revealed some melanin granules. Migration took place towards the defect in the RPE layer. At the edge of the lesion Bruch's membrane was wrinkled and choriocapillary vessels were closed.

After 7 days, proliferation could be clearly demonstrated. The whole defect in the RPE layer due to the laser impact was already covered by a mound of focally nodular, multilayered RPE. Maximal proliferation was found predominantly at the edge of the lesion. In this area the melanin granules lost their apical orientation. Bruch's membrane was strongly wrinkled in that area. The choriocapillary vessels were open at the edge of the burn. The whole lesion was estimated to be 340 Ìm in diameter, but proliferation made it difficult to determine the exact dimension of the lesion.

Hence due to the cell proliferation of untreated cultured RPE it was also difficult to determine the diameter of the laser lesions 7 days after cultivation. After 14 days, cells with broadly distributed melanin granules were present. Multilayered cells showed some prominent proliferations. The diameter of the lesion was not determinable because of the general cell proliferation but Bruch's membrane which was focally wrinkled again, was indicating the presence of a lesion.

Control Experiment

The permanent perfused culture model offers the possibility to cultivate retinal tissue in an organotypical environment having a perfusion line from both sides, the upper RPE and the lower choroid side. Using this RPE organ model in perfusion culture, it was possible to examine the behaviour of the RPE after laser photocoagulation and the behaviour of the organ culture without treatment.

The proliferative behaviour of the RPE is associated with the substratum on which the cells are cultured. We used the perfusion culture system with the RPE monolayer still attached to Bruch's membrane. In previous static culture approaches it was difficult to maintain the RPE as an intact monolayer. Numerous workers developed culture models but had difficulties finding a solid support for the exoplants of human RPE and choroid.

Tso et al. used filter paper (Millipore) as support for the isolated uveal tissue and noted a markedly wrinkled Bruch's membrane and hypopigmented RPE cells after only 3 days in organ culture and frequent focal nodular proliferation within 10 days of culture.

In contrast Del Priore and coworkers did not observe these alterations using another model. They were able to maintain the RPE as a monolayer for some weeks and explained this by the fact that the RPE and choroid remained attached to the structural support of the underlying sclera in their organ culture system. In their experiments the RPE monolayer remained intact after 2 weeks but they also noticed hypertrophic RPE cells and areas of RPE hyperplasia.

Our approach using the permanent perfused culture system supports the findings of other authors, who recognized proliferation of the cells. It was speculated that with this new culture model, RPE without neurosensory retina should be maintained as an intact monolayer for longer than 2 weeks. After 3 days in the perfusion culture the RPE monolayer was intact, after 1 week the cells were also intact as a complete monolayer, though there were some morphologic changes. However, after 2 weeks a marked proliferation of the cells and multilayered RPE was observed.

Proliferation of the cells as seen beyond 7 days could be interpreted as a sign supporting the vitality of the organ culture. However, the morphologic findings after 14 days of culture may also suggest a phenotypical change of the RPE cells into other cells such as macrophages. Alternatively an overgrowth of the cell layer due to choroidal cells seems to be unlikely since Bruch's membrane stayed intact.

In our study these cell alterations of untreated RPE limits the use of the culture system for RPE organ culture over a period longer than 2 weeks. It is not clear why this cell change occurred in the RPE culture. When full thickness adult retina is cultured in the perfusion system, no signs of cell changes are observed, and the RPE remains intact as a monolayer over a period of 4 weeks. Therefore it could be speculated that with the loss of neurosensory retina the RPE is no longer inhibited, and thus reacts with cell proliferation and subsequently further scarring.

In comparison to the static culture approach as described by Del Priore the perfusion culture system revealed no advantage for RPE cultures since proliferation of untreated cells was evident after 7 days of culture, whereas Del Priore was able to maintain the RPE for 28 days.

Laser Application

The effects of laser treatment of the fundus have been studied by several groups. In vivo it has been observed that argon laser photocoagulation of monkey and human fundus causes necrosis of the RPE and a lifting of the RPE from Bruch's membrane, budding of individual RPE cells and multilayered RPE formation in the area of laser irradiation by 7 days after treatment. Histologic sections revealed that by irradiating the RPE with a conventional argon laser the whole area of the cells was destroyed and the choriocapillaris as well as the choroidal vessels were damaged.

After laser photocoagulation RPE cells migrate and proliferate to cover the defect. In vivo after mild photocoagulation (just light whitening of the irradiated area), as usually performed in macular coagulation, the RPE barrier becomes intact again. Histological examinations of rabbit retina showed also a proliferation of the RPE after coagulation with short pulsed repetitive laser systems without destruction of the neurosensory retinal layer. So RPE proliferation can always be observed as a healing mechanism to repair any defects in the RPE monolayer or adjacent structures.

The effects described could also be demonstrated in the laser lesions of our approach investigating mild RPE laser photocoagulation in vitro. In these examples, as in vivo, RPE cells started to cover the defect by migration and proliferation.

After one week the defect was completely covered by multilayered RPE cells showing still an ongoing proliferation after 14 days. Similar to the findings of Del Priore, one day after argon laser photocoagulation we could also observe a release of the RPE from Bruch's membrane and a total destruction of the irradiated area. It was interesting to note that Bruch's membrane itself was not destroyed but appeared wrinkled.

Thus it was possible to show in vitro the migration and proliferation of the neighboring RPE cells to cover the laser defect up to 7 days. This is consistent with the in vivo findings of the RPE selective laser treatment showing that the RPE is replacing the irradiated area.

These healing mechanisms occur even without overlying neurosensory retina. This is an important observation because the neurosensory retina contains potent growth factors that could affect the morphologic response of the RPE to laser photocoagulation. However, since this is not an audio-radiographic study, it should be noted that the 'proliferation' might just be a metaplasia rather than a mitosis of the cells.

Limitations of the Study

Most authors did not give information about the number of laser spots examined, sizes of lesions in histology and percentage of proliferations. In our preliminary experiments the described effects of laser-tissue interaction could be observed in a minority of all applied laser spots.

This could be mainly explained by the difficulties in identifying the laser burns when trimming the tissue due to the fact that no thermal necrosis could be seen because of the lack of neurosensory tissue. Thus major damage is only confined to the RPE and in spite of the macroscopically greyish appearance of the RPE in irradiated areas histological alterations are hard to identify. Concerning acute laser lesions Bruch's membrane is denuded and void of RPE cells.

In our study several areas without RPE were found – also 3 and more days after irradiation – but differentiation from artifacts due to tissue preparation was difficult and therefore no fraction of proliferating lesions could be mentioned. In contrast to this, localized proliferations in treated specimens were a clear sign for laser lesions.

Another explanation for the irregularity of present proliferations could be the possible variation in laser energy, since it turned out to be difficult to judge experimentally an endpoint of laser treatment, when neural retina is absent. However, in all treated specimens laser burns were slightly visible indicating a destruction of the RPE which should be independent from the delivered energy.

Also the variation in spot size due to different focussing could be a contributing factor for the different proliferation behaviour of the RPE. Interestingly the described laser lesions showed variable sizes of destruction. Although the chosen diameter of irradiation was about 200 Ìm, the diameter of the histologically observed laser lesions varied. There are several explanations for these findings.

The focus during laser application through the microscopic culture chamber could change the diameter and additionally, each burn was adjusted and focused individually with a certain variation. Secondly, RPE cells responded variably to laser coagulation depending on their melanin content. This would lead to a variation in heat conduction. After 7 days it was nearly impossible to determine the exact size of the lesion because of the overall proliferation.

A major limitation for studying laser-tissue interaction for periods longer than 14 days seems to be the perfusion culture system itself because both treated and untreated tissue revealed proliferation and thus differentiation and demarcation of treatment results are difficult. Therefore, further research should consider improving the perfusion system, e.g. by changing the buffer solution or the perfusion time. This model is useful to study laser effects of the RPE in vitro although preparation and histological documentation of laser spots can be difficult.

However a healing process of the RPE could be observed associated with migration and proliferation of the cells. On the other hand, untreated RPE also showed a proliferation after 7 days in the perfusion culture. After 2 weeks a general alteration of RPE cells to multilayered cells occurred. This morphogenesis of cultured RPE cells has been already shown by several authors.

As a consequence of these RPE alterations our system seems not to be useful after a period of 7 days to examine laser tissue interaction. In contrast, it might be suitable – because of short-term changes – to perform e.g. toxicity tests with intraoperatively used balanced solutions or investigations on growth factor production or RPE transplantation in the early period.

Index

A

Acrylamide 263
Aids Denialism 204
Airborne 31, 150, 186, 187, 188, 189, 190, 192
Algae 1, 23, 24, 27, 34, 37, 38, 42, 43, 56, 57, 76, 152, 153, 154, 157, 158, 159, 238
Alternative Medicine 214
Antibiotics 39, 41, 186, 191, 193, 198, 261, 262, 263, 264, 265, 266, 288
Antimetabolites 221, 224, 226, 229, 232, 234
Antimetabolites 221, 222
Antiviral Therapy 212
Ascorbic Acid 225, 229, 230, 231, 232, 233, 242
Aseptic Technique 44
Asthma 175, 176, 177, 178

B

Bacteria 4, 14, 23, 24, 25, 27, 28, 34, 36, 37, 38, 40, 41, 42, 43, 73, 75, 87, 88, 119, 159, 160, 197, 222, 224, 251, 265, 266, 282, 283, 284, 288, 289, 290
Biochemical Tests 158
Biogeochemical 285, 287
Bioinformatics 103, 104
Biological 1, 2, 3, 4, 6, 8, 24, 41, 54, 55, 58, 59, 62, 77, 78, 79, 104, 119, 125, 130, 131, 135, 226, 228, 230, 233, 234, 239, 271, 272, 276, 277, 278, 279, 280, 281, 292
Biomarkers 286
Biomolecules 84, 85, 268
Biosynthesis 82, 111, 113, 114, 119, 120, 228
Biotechnology 267, 268, 269, 270, 271, 273, 274, 275, 276, 277, 278, 279, 280, 281
Bioterrorism Agents 183
Biotin 80, 222, 223, 224
Blood-borne Pathogens 203

C

Candidiasis 191
Carcinoma 191
Cells Affected 206
Cellulose 107, 108, 109, 110, 111, 112, 113, 114, 116, 117, 118, 125, 126, 127, 129, 131, 132, 262

Cervix 187, 191
Chancroid 186, 202
Cheese Manufacture 249, 250, 251
Cheese Ripening 249, 251, 257, 258
Chemical Changes 33, 34
Chemistry 1, 12, 16, 17, 25, 64
Chlamydia 189, 190, 192
Choline 234, 235, 236
Chronic 175, 176, 177, 192, 198, 205, 207, 217
Clearance and Immunity 160
Common Cold 40, 169
Common Respiratory 175
Current Progress 134, 136
Cyanocobalamin 227, 228, 229

D

Dairy Enzymes 259
Dairy Industry 249, 254
Dairy Processing 248, 249, 260
Dark Field 59, 71, 73
Deconvolution 62, 63
Deficiency Syndrome 225, 228, 231, 236
Dilution Plating 50
Disease Transmission 164, 165, 166, 168, 174
Disorders 81, 175, 180, 181, 223, 292
Droplet-transmitted Diseases 167
Dry Heat 54
Dyspnea 175, 176, 177, 181

E

Ecological 278, 285, 286, 287
Electron Microscopy 14, 16
Environmental 135, 165, 267, 269, 270, 271, 272, 273, 274, 275, 276, 277, 278, 279, 280, 281, 284
Enzymes 14, 39, 119, 123, 124, 131, 159, 223, 248, 249, 251, 253, 255, 256, 257, 258, 259, 260, 261, 262, 263, 268, 271, 275
Epidemiology 152, 186, 215
Epididymo-orchitis 193
Ethical 134, 136, 144, 146, 148, 149
Exposure 19, 54, 55, 88, 100, 128, 151, 165, 170, 173, 196, 202, 203, 205, 209, 210, 211, 212, 239, 244, 245, 266
Eyepiece Separation 74

F

Fabrication 89, 95, 98, 104, 134
Fat-soluble Vitamins 237
Fermented Milk 258, 259
Fibrosis 175, 181
Field Microscopy 57, 58, 59, 75
Flame Sterilization 44, 45
Fluids 211
Fluorescence Microscopy 61, 62
Folic Acid 224, 225, 226, 227, 228, 229, 230, 236
Fungi 23, 24, 27, 34, 38, 39, 40, 42, 43, 52, 56, 57, 88, 150, 151, 197, 252, 266
Future Research 213

G

Gastrointestinal Infections 198
Gene 13, 16, 61, 81, 82, 85, 95, 96, 98, 100, 102, 103, 105, 106,

137, 143, 182, 205, 251, 252, 253, 265, 269
Genital Warts 190
Gonorrhoea 186, 194, 195
Grind 5

H

Hemicelluloses 107, 108, 109, 110, 113, 114, 115, 116, 117, 118, 119, 121, 122, 123, 124, 125, 126, 127, 129, 130, 131, 132, 133
Herpes Genitalis 190
Hierarchic Organization 109, 132

I

Idiopathic 175, 181
Immune System 43, 44, 89, 90, 91, 92, 154, 195, 197, 199, 201, 204, 205, 206, 207, 216
Immunology 91
Inactivation 221, 224, 229, 232
Industrial Production 264
Infected Body 210, 211
Infectious Disease 26, 42, 150, 151, 152, 154, 156, 157, 159, 161
Influenza 163, 166, 169, 170
Inositol 233, 234, 235
Intensive Care 183
Intra-molecular Interactions 80
Isotopes 285, 286

L

Lactase 254, 255, 256, 261
Lactic Acid 25, 254, 258, 259
Laser Application 291, 295, 296, 299
Laser Scanning 62
Lens 24, 27, 28, 57, 64, 66, 69, 71, 72, 73, 74, 76, 293
Light Microscopes 57, 66, 70, 71
Lignin 107, 108, 109, 110, 111, 112, 113, 114, 116, 117, 118, 120, 122, 123, 126, 127, 129, 130, 131, 132
Lipoic Acid 237
Lymphogranuloma 190

M

Malignancies 199, 200, 216
Marine Creature 10
Mechanics 8, 81
Microarrays 90, 95, 96, 98, 99, 100, 102, 103, 105
Microbes 13, 24, 27, 30, 31, 32, 33, 34, 35, 38, 40, 41, 42, 43, 44, 46, 47, 49, 151, 152, 157, 158, 163, 164, 246, 247, 248, 270, 271
Microbial Biomass 281, 284, 289
Microbial Cultivation 51
Microbial Culture 156
Microbial Enzymes 257, 261
Microbial Nitrogen 283, 284, 288
Microbial Rennets 250, 258, 261
Microbiological 156, 226
Microscopes 24, 27, 28, 34, 56, 57, 59, 62, 64, 66, 70, 71, 72, 73, 75
Misconceptions 204
Misfolding 15, 18, 21
Modes of Transmission 186
Molecular Diagnostics 155, 159
Mortality 163, 170, 172, 178, 179, 181, 183, 184, 185, 196, 213, 214, 215, 218
Mother-to-child 212

N

Nacre Crystalline 21
Nanobiology 1, 2, 3, 17, 77, 86, 107
Nanomachines 3
Neurological Diseases 199
Niacin 219, 220, 221, 222
Nicotinic Acid 219
Nitrogen Cycle 281, 282, 284
Nonmedical Use 266

O

Obstructive Pulmonary 175, 178
Oligonucleotide Arrays 99, 100
Opportunistic Infections 195, 197, 198, 200, 204, 207, 214, 216
Optical Microscopy 57, 58, 59

P

Pandemics 162, 163
Pathophysiology 204
Peering 11
Peptidases 257
Perinatal 203, 210
Phase Contrast 59, 60, 71
Positive Functions 219, 223, 234, 238, 240, 242, 244
Precursors 108, 110, 112, 114, 115, 118, 120, 122, 123, 124, 125, 130, 144, 262
Preventing Transmission 153
Prevention 174, 192, 193, 196, 208, 209, 211, 216
Probe 12, 14, 16, 78, 80, 81, 87, 96, 99, 100, 103, 106
Prognosis 184, 189, 200, 215
Proposition 135, 136, 146, 147, 148
Protection 43, 161, 194, 202, 217, 226, 234, 239, 241, 243, 245
Protein 2, 6, 8, 10, 13, 15, 18, 19, 21, 25, 61, 62, 77, 78, 81, 82, 83, 85, 86, 87, 88, 90, 92, 159, 182, 220, 221, 222, 223, 227, 228, 238, 250, 252, 254, 258, 259, 280, 289
Proteinases 249, 250, 257, 258, 259, 261
Proteolytic Enzymes 258
Pulmonary 167, 175, 176, 177, 180, 181, 182, 183, 184, 198, 208
Pulmonary Diseases 177
Pulmonary Infections 197

R

Recombinant Rennets 251
Risk of Transmission 165

S

Sars 166, 167, 168, 174
Sensor 77, 78
Sexual Transmission 190, 192, 193, 210, 215
Sexually 186, 187, 188, 190, 191, 193, 195, 202, 203, 208, 210
Side Effects 89, 212, 213, 214, 266
Sightless 10
Simulating Life 12
Sleep Apnea 175, 180, 181
Slick Joint 6
Specimen 58, 59, 60, 66, 67, 71, 72, 73, 74, 75, 208, 291, 293, 294, 295, 296
Spontaneous 29, 30, 31, 80, 141
Spotted 96, 99, 100
Staging System 207

Staining 58, 67, 68, 158
Standardization 104, 138
Statistical Analysis 103, 105
Stem Cell 134, 136, 138, 139, 142, 144, 145, 146, 147, 148, 149
Sterilization 32, 33, 44, 45, 53, 54, 55, 56
Stigma 193, 217
Strains 153, 169, 201, 203, 250, 261, 265, 266
Streak Plates 44, 47, 48
Sub-diffraction 63, 64
Surface Engineering 98, 99
Syndrome 7, 166, 184, 187, 188, 190, 195, 220, 221, 223, 234, 238
Syphilis 66, 152, 186, 187, 188, 189, 194, 195, 202

T

Tertiary Structure 83, 114, 126
Therapeutics 88, 249
Thromboembolism 182, 183
Tissue Engineering 12, 134, 135, 136, 137, 138, 141, 145, 148
Tissue Preparation 301
Transmission 67, 150, 152, 153, 154, 155, 164, 165, 166, 167, 174, 187, 190, 192, 193, 196, 199, 202, 203, 204, 209, 210, 211, 215, 235
Treatment 7, 26, 32, 33, 40, 56, 88, 103, 104, 105, 135, 137, 141, 142, 145, 154, 155, 175, 266, 269, 271, 278, 279, 281, 299, 300, 301
Trichomonas Vaginalis 192
Tube Transfer 44, 45
Tuberculosis 41, 152, 166, 167, 174, 176, 201, 208
Tumors 88, 89, 177, 195, 200

U

Urethritis 189, 190

V

Vector-borne 150, 164, 172
Venereum 190
Viral Hepatitis 191
Viscoelastic 84

W

Wet Heat 53, 54
Wheezing 175, 177, 179

X

Xanes 17, 18, 19, 20, 21, 22

Z

Zoonotic Diseases 172